W0259642

The Flower Thesaurus

1

BBC Books, an imprint of Ebury Publishing
Penguin Random House UK
One Embassy Gardens, 8 Viaduct Gdns,
Nine Elms, London SW11 7BW

BBC Books is part of the Penguin Random House group of companies whose addresses can be found at global.penguinrandomhouse.com

for full copyright details please see p.282

First published by BBC Books in 2025

www.penguin.co.uk

A CIP catalogue record for this book is available from the British Library

ISBN 9781785949258
Commissioning Editor: Nell Warner
Editor: Phoebe Lindsley
Design: Double Slice Studio | Amelia Leuzzi & Bonnie Eichelberger
Copyeditor: Kay Halsey

BBC Gardeners' World Magazine: Kevin Smith, Sarah Edwards

Compiled by *BBC Gardeners' World Magazine*

Printed and bound in China by C&C Offset Printing Co., Ltd

The authorised representative in the EEA is Penguin Random House Ireland, Morrison Chambers, 32 Nassau Street, Dublin D02 YH68.

Penguin Random House is committed to a sustainable future for our business, our readers and our planet. This book is made from Forest Stewardship Council® certified paper.

The Flower Thesaurus

Colourful Pairings and Plant Ideas for the Creative Gardener

Liz Potter

Contents

Introduction

How to use colour in the garden

Gardeners are often driven by the pursuit of 'colour', but flowers are so much more than an array of pretty petals and stamens. They are, to put it bluntly, the sex organs of the *angiosperms* (flowering and fruiting plants), which evolved around 130 million years ago, when dinosaurs still roamed the earth. In fact, flowers have played a pivotal role in evolution – enriching biodiversity once the dinosaurs became extinct (around 66 million years ago). Flowering plants came to dominate life on land, creating vast rainforests that supported millions of insects, lizards, mammals and birds.

Flowering plants remain an integral part of our ecosystem – human life depends on them for the vast majority of our food. When fertilised by the transfer of pollen, the ovary behind the flower typically swells into a fruit containing seeds that the plant uses to reproduce. Most of our agricultural produce – grains, seeds, fruit and vegetables – are made in this way.

So, in terms of our own planetary survival, it's just as well humans are happy to cultivate flowers. Gardeners are notoriously 'flower greedy', enticed by their myriad colours and intricate shapes, and spending much of the year eagerly anticipating the next bloom, and the next. Feelings of profound gratitude can swamp our sensitive souls in springtime as each plump little bud unfurls to reveal a perfect blossom that has never seen daylight before.

We know instinctively that colour plays a significant role in our floraphilia, but scientific studies have also shown that humans prefer certain colours because we associate them with positive emotions, often related to the natural world. For instance, many of us like blue because it reminds us of clear summer skies or the seaside, and so on. Seeing favourite flower colours can promote the release of dopamine, the happiness hormone, and increase general wellbeing.

Although flowers often take centre stage in our affections, foliage colour can be just as engaging, lasting much longer than the fleeting, seasonal blooms they partner. Leaves can be white, black, yellow, purple, silver, bronze, red, orange or blue; striped, splashed or flecked. They can make or break a colour scheme and are arguably more important than the flowers themselves.

Colour can be so dazzling in the garden that it's easy to forget that it's but one component of any successful planting matrix: there are shapes, textures, sounds, tastes and fragrance to explore too. If you only focus on colour, there's a danger you'll create a one-dimensional experience – a garden that has plenty of 'wow' factor but little else to engage your other senses. In this book, we'll be exploring the many plant colours available in the gardener's palette, but also highlighting plants that will additionally offer interesting shape, form, texture and fragrance across the year. This way, you'll create a satisfying symphony of plants, rather than a 'one-hit wonder'.

Design considerations

Alas, creating a garden is not quite as simple as picking your favourite flower colours at the garden centre and hoping for the best. Although many of us do shop this way – flowers are so seductive it takes strength to resist them! – quite often this approach lets us down, and the plants die off unexpectedly. This is not only disappointing, it's expensive and wasteful.

For a garden to be self-sustaining in the longer term, we need to choose plants that will thrive there. This means considering the garden's size, aspect, site and soil.

We need to ask:

1. How much space is there for planting, and how do I want to divide the area among the plants? Is there space for a small tree or large shrub? Is the lawn too big? Do I need a lawn at all?
2. How does the sun track across the sky from dawn to dusk? Where are the main pockets of shade and sunshine during the day and across the seasons?
3. Is the soil rock hard in summer and slippery wet in winter rains, or is it sandy and free-draining, so plants often go thirsty in summer?

It's important to understand the garden first, then choose 'the right plants for the right place' within it. We need to shop for the real-life conditions of our plot, otherwise new plants will simply sulk, or die.

It helps to look around the neighbourhood to see what's thriving for clues about the type of soil we have, and to invest in a cheap soil-testing kit from the garden centre. This will help measure the soil's pH (potential for hydrogen), which indicates whether it's acid, alkaline or somewhere in between. Ericaceous plants, such as many heathers and rhododendrons, prefer acidic soils; lime-loving campanulas prefer alkaline ones.

If you're faced with a newbuild garden that's full of builders' rubble lurking under the surface, take heart. It's entirely possible to improve your soil for planting: remove any debris first, then buy in fresh topsoil by the ton bag. Layer on organic matter as an annual mulch, create your own compost bin and save any fallen leaves to make nutritious leafmould soil conditioner. All these activities will feed the soil and create a happier, healthier root environment for plants.

In a larger garden, it may help to map out the garden's areas of full sun and deep shade, clay and sand, boggy hollows and dry patches.

We also have to consider our garden's location, its 'borrowed views' (eg neighbouring trees), eyesores to hide, and the design concept you're going for. Colour is but a small part of this jigsaw, but one that's undeniably very exciting and motivating.

Plan for seasonal changes

Next, we need to consider how the seasons will progress in the garden. In particular, choose a few plants that will offer structure and colour in autumn and winter; this is a gloomy time of year and, all too often, if you fill the garden with pretty summer annuals and herbaceous perennials that die back in autumn, the garden will revert to 'mud and sticks' come wintertime. That's why it's important to add in a few evergreen shrubs and trees whose glossy leaves will shine all year round, plus a few deciduous shrubs for their spring or summer blossom, vibrant autumn leaf colour, fruit and berries.

Spring and summer are relatively easy to plant for, but it's worth thinking of summer as a game of two halves best described as 'early summer' (May to July) and 'late summer' (August to September).

People will often tell you that gardening is an art form, like painting a watercolour, but this is a grave understatement of the gardener's skill and foresight! In fact, creating an attractive garden is exponentially more difficult – rather like painting in *four* dimensions, not just two. For one thing, the garden is a three-dimensional, immersive space that, when planted thoughtfully, can appeal to all the senses with layer upon layer of colour and interest. But then you also need to factor in the fourth dimension: Time. How will the planting scheme evolve across the seasons, and through the years ahead?

Across a single year, foliage and buds will change colour and flowers will come and go. Plants will grow in size and perhaps become too big for their space, leaning on or even engulfing their neighbours. They can produce attractive seedheads and berries in strange new shapes and colours, perhaps joined by their own self-sown babies. Some plants can be made to produce more flowers for longer, by deadheading spent blooms with the secateurs. In winter, plants can retreat below ground or shrug off all their leaves to survive the cold weather, only to return in spring. A successful planting scheme needs to take all these potential changes into consideration.

Using the colour wheel to organize your choices

Theories on colour go back as far as Aristotle, but it was Isaac Newton who discovered that white light could be split into a visible spectrum of colours, and pioneered the use of the colour wheel in the early 1660s. In his model, he presented seven colours to match the number of notes on a music scale. Since then, artists, philosophers, scientists and writers have all explored how colour is received by our eyes, and how pigments can be mixed together to create different paint hues, shades and tints.

The modern colour wheel is a useful tool for organising plants in the garden, too. It's a device that helps you see instantly the colours that will harmonise well together or make an attractive contrast. The artistically inclined gardener might like to paint their own wheel on a sheet of paper: you need just the three primary colours of red, blue and yellow.

Mix these primaries to create the three secondary colours: green, orange and violet. You can then create the six tertiary colours by mixing the secondaries with the primaries, to make red-orange and blue-violet etc. Using just these twelve basic colours, plus white, black and grey (in theory) you can create all the colours you'll ever find in nature.

Playing with colours is enormous fun, but to narrow things down and work out your own preferred colour schemes in the garden, there are a few tried-and-tested approaches you can start with:

HARMONIOUS COLOUR SCHEMES

Include any three to five colours that sit next to each other, adjacent on the colour wheel, such as red, red-orange and orange or yellow-orange, orange and yellow. Usually, one of those colours is dominant, while the others are brought in as background players. Both hot and cool colours can be used to create harmonious effects – but it's best to avoid mixing cool and hot tones, for example, shrubs that are gold (sunshine) with silver (moonlight).

COMPLEMENTARY SCHEMES

These are the contrasting hues that lie opposite each other on the colour wheel. The most commonly seen flower pairings are purple and yellow, red and green, and blue and orange. These colour combinations create a more dynamic effect with a punchier feel than harmonious shades. They are colours to lift your mood and work well with some light foliage woven through, adding a little visual 'downtime' to the scheme. Don't just think of flowers – coloured leaves can also be used as one half of a complementary scheme; perhaps even consider a scheme using foliage alone.

MONOCHROMATIC SCHEMES

Use only one hue, with light and dark variations to add interest. White and neutral green are often considered as monochromatic partners, leading to an elegant, minimalist scheme where nothing jars.

CLASHING SCHEMES

These tend to be bold and courageous, assaulting the senses with strong colours that lie adjacent on the colour wheel, but that fight against harmony. For instance, orange and pink are relatively combative bedfellows. Many cornflower meadows rely on this sensory assault too. To tone down a clashing scheme, thread a neutral foliage colour, such as green, through the planting.

COLOUR ACCENT SCHEMES

These are a handy way to add 'oomph' to a harmonious or otherwise monochromatic scheme. By dropping a single blue or purple plant into a sea of yellow or orange you'll create a vibrant 'look at me' statement that suggests a strong degree of colour confidence.

How to use this book

Over the following pages we'll take you on a guided tour of our favourite colourful plants – both foliage and flowers – to help you achieve exciting colour effects in your garden. Each chapter focuses on one of the key colours of the visible spectrum[1], and comprises a team of fifteen to twenty hero plants[2] that are both reliable and versatile, suitable for most garden types in either full sun, part shade or full shade.

All of our hero plants will look good together in a colour-themed bed or border on a mingling, mix-n-match basis. They're also accompanied by a choice of three planting partners[3] that will either match, harmonise or contrast with them, or provide exciting foliage, fruit or stems alongside. You don't have to use all three partners with the hero plant, but you can be confident they'll rub shoulders attractively.

The hero plants are organised in chronological order by their month of flowering, starting with spring. This way you can get the seasonal timings right for a succession of year-round colour, with something of interest to draw you outdoors, even in the depths of winter.

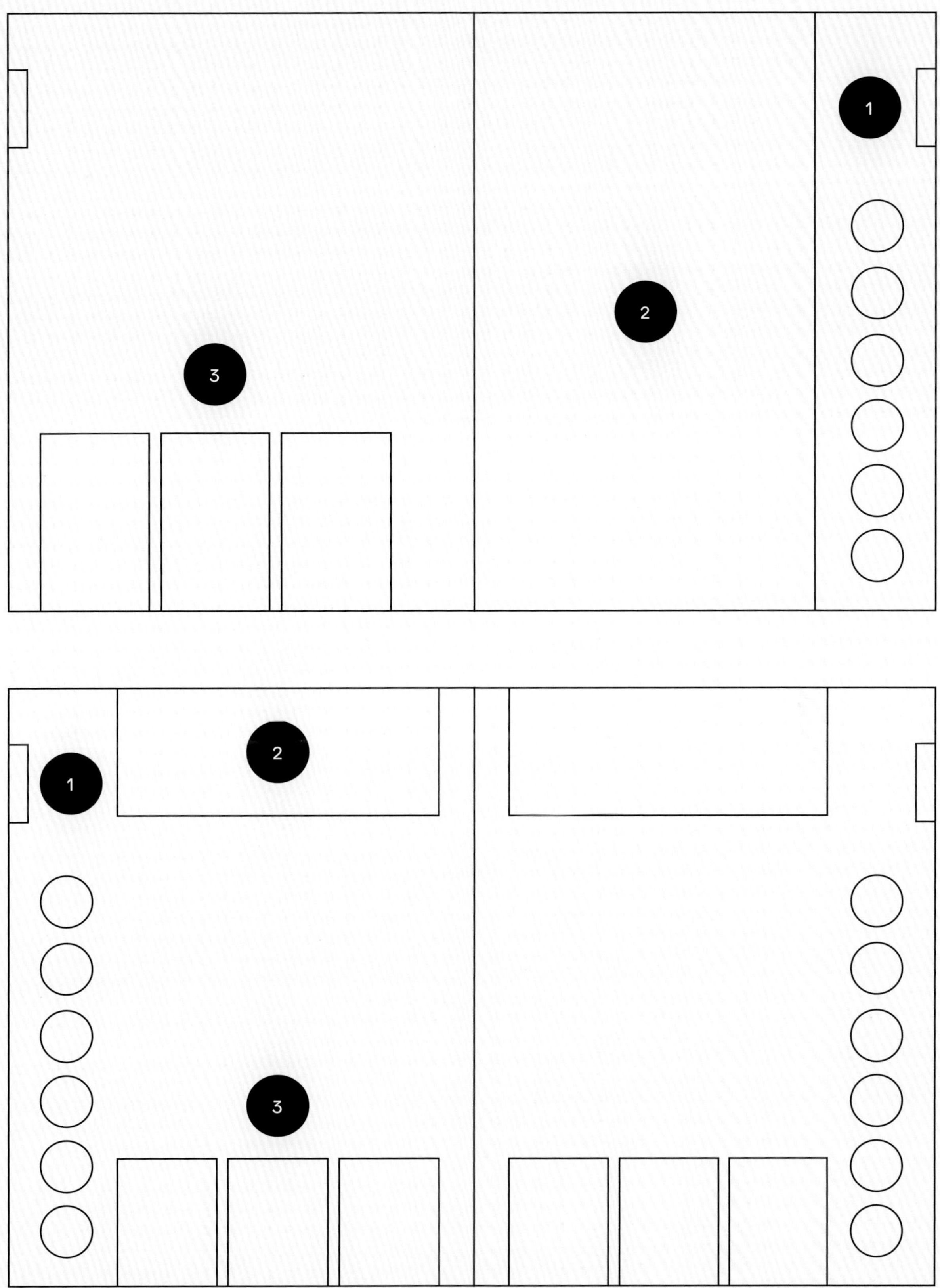
1
2
3
2
1
3

Red

RED is the colour of passion: fire and danger, heat and blood. It is Temptation and Hellfire, hot chillies, romantic roses and sun-ripened tomatoes. This mix of cultural connotations might lead us to treat it with caution in the garden – you don't need much to make a statement – and yet there are so many exciting plants with red flowers and/or purple-red foliage that it would be a shame to deprive yourself.

Think of how the artist JMW Turner added a little last-minute daub of red paint – a bobbing buoy – into his cool-grey seascape *Helvoetsluys, The City of Utrecht, 64, Going to Sea* (1832) in order to steal attention from a neighbouring painting by John Constable at the Royal Academy ... That little buoy ignited a feud: it's powerful stuff!

A complicating element for gardeners lies in the fact that there are lots of different reds to choose from. There are orange-reds (eg poppies, tomatoes), pink-reds (eg astrantias, peonies) and blue-reds (eg roses, penstemons) and they don't always sit well together. If in doubt, always buy red-flowered plants in person, in bloom, and if you want to match them together, buy both plants at the same time so you can see how they look in your shopping cart before taking them to the till.

In the spring garden, purple-red hellebores and ruby-red tulips make a vivid contrast with all the seasonal yellows and blues. At this time of year, there are few natural reds about – save for the festive chest of your garden robin – so those reds that do occur are all the more precious. Shrubs such as red-flowering *Chaenomeles* are particularly eye-catching, the flowers emerging on bare twiggy stems, and draw in early pollinators on the wing.

May and June bring red poppies, peonies and roses. These blowsy blooms are perfect for a cottage-garden look, with plump buds that burst open to cup-shaped, bee-friendly flowers and petals that evoke sensuous ballgown fabric. Perennial *Papaver orientale* poppies emerge from a furry green casing that splits open to reveal a crumpled bundle of petals, like a silk prom dress emerging from a too-small suitcase. Herbaceous peonies are almost velvet by comparison, the single-flowered types often with a centre of contrasting golden petals or stamens. Their shoots emerge from the soil like crimson knuckles in spring. Rose petals have a more satin-like texture, prized for potpourri and wedding confetti; these romantic flowers unfurl from June onwards, often exuding a heady fragrance.

By late summer there's an opportunity to ramp things up by creating a vivacious, exotic-looking 'hot border'. These dramatic colour-themed planting schemes are best located in a warm, sheltered, sunny part of the garden, where the vibrant scarlet flowers of crocosmia, dahlias and canna lilies will mingle with bruised, burgundy foliage and blushing grasses. A famous example can be found at Hidcote Manor in the Cotswolds, said to be inspired by the various shades of red in the folds of a dress worn by solo cellist Madame Suggia in a painting (1920–3) by Augustus John.

Red dahlias are a mainstay of any hot border and there are plenty to choose from – dark-leaved, single-flowered 'Bishop of Llandaff' is a favourite, but there are lots of eye-catching alternatives, such as cactus-type 'Doris Day' and waterlily-type 'Witteman's Superba'. To pile on the drama, pair them with darker red dahlias such as almost-black 'Nuit D'Ete' (syn. 'Summer Night') and semi-dinner-plate 'Arabian Night'. They both have oodles of exotic charm that's perfect for creating visual fireworks in your red-themed hot border.

By the time autumn rolls around and your hot border starts to fade, be sure to have a vibrant red Japanese maple waiting in the wings. There are lots of red and purple-leaved options, such as *Acer palmatum* 'Atropurpureum', 'Bloodgood' and 'Garnet'. These slow-growing small trees, with lobed palmate or finely dissected filigree foliage, can provide a glorious crescendo of colour in a sheltered shady spot and look perfect in a Japanese-inspired courtyard.

As autumn cedes into winter, we begin to rely on red stems and bark for our garden drama. When deciduous dogwoods lose their leaves they reveal a handsome skeleton of colourful stems – *Cornus alba* 'Sibirica' being the best for pure red. Depending on how you prune them (ideally once a year in late February or early March, back to approx. 10cm) you can create a lovely dense thicket of fiery stems seeming to rise up from the ground like a flamboyant bonfire. They look especially dazzling planted where they will catch the low, raking winter sun.

In the depths of winter, wild birds will seek out red in the form of delicious crimson berries and crab apples. There are numerous red-berried hollies to choose from – such as *Ilex aquifolium*, which also comes in a variegated form, *Ilex aquifolium* 'Argentea Marginata', with leaves margined in silver-cream. Another popular option is *Ilex x altaclerensis* 'Golden King', a gold-variegated female tree (despite its male name) that's almost spine-free. Buy a female cultivar if you want berries but note that they will rely on a male holly nearby for fertilisation.

In winter especially, red is a colour that can instantly revivify garden borders that are otherwise quite dull and lifeless. As Christian Dior once said: 'There's a red for everyone.' That's certainly true when it comes to red plants. Whether you want to create a sizzling 'hot border' or raise the tempo in a cool sea of blues, there's sure to be a red plant just for you.

Helleborus hybridus

Nodding, long-lasting oriental hellebores are the mainstay of any part-shady garden in winter, flowering February to April, with attractive evergreen palmate foliage. Plant them in threes or fives under a deciduous shrub or small tree for a woodland-edge effect, or under an evergreen that gets a little sunlight in winter. They prefer humus-rich soil, so it's a good idea to add a layer of organic mulch in spring.

PLANT WITH

Bergenia cordifolia

The leathery evergreen foliage of bergenia provides year-round ground cover in shade. The pink flowers appear in March and April to partner blushing pink-red hellebores perfectly.

H60cm x S75cm

Helleborus foetidus

From the same hellebore genus, this 'stinking hellebore' is more delightful than its name suggests. Cup-shaped green flowers have a purple edge, flowering January to April.

H80cm x S45cm

Cornus alba 'Sibirica'

Plant your hellebores among the naked red stems of this striking dogwood for an attractive winter cameo. Attractive acid-green buds will emerge in March – unless you cut the stems back.

H2.5m x S2.5m

H30cm x
S45cm

Sun or
Part Shade

Moist Well
Drained Soil

Pollinators

Frost Hardy

Tulipa
'DOLL'S MINUET'

Lily-flowered, magenta-red viridiflora tulip flushed with a green stripe, flowering April to May. Broad, lance-shaped leaves. Plant tulip bulbs in November, about twice as deep as their length (approx. 8-10cm), pointy end upwards. Tulips prefer a light, well-drained soil in sun or part shade for best chance of a repeat performance next year.

PLANT WITH

Erysimum cheiri
'Red Jep'

Semi-evergreen perennial with flowers in red, reddish-purple and pinky-orange. Best in full sun and an excellent partner for any red tulip.

H50cm x S50cm

Myosotis sylvatica

Blue or white forget-me-nots make excellent partners for tulips, helping to hide tatty foliage and creating a sea of colour, April to June.

H30cm x S15cm.

Veronica
'Heartbreaker'

Evergreen hebe that blushes pink as temperatures drop, forming neat mounds in spring and summer. Purple flowers June to August.

H60cm x S60cm

H55cm x
S15cm

Sun or Part
Shade

Moist Well
Drained Soil

Frost Hardy

Peony
'SCARLET HEAVEN'

An intersectional hybrid or Itoh peony, flowering May to June, with large single flowers that have a ring of yellow stamens inside and finely cut, dark-green leaves tinged bronzy-red in spring and autumn. Plant in a sheltered sunny or part shady spot in a rich but well-drained fertile soil, mulching with well-rotted compost or farmyard manure.

PLANT WITH

Astrantia
'Claret'

These dark burgundy-red perennials make a dainty contrast for bigger peony flowers, flowering June to August.

H90cm x S30cm

Salvia x sylvestris
'Mainacht'

The upright purple spires of *Salvia x sylvestris* 'Mainacht' flourish in sun or dappled shade and form a vibrant contrast with red peonies.

H75cm x S45cm

Foeniculum vulgare
'Purpureum'

Bronze fennel offers an attractive feathery foil for peonies and astrantias.

H1.8m x S45cm

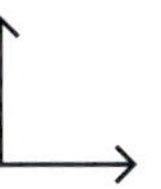

H75cm x
S50cm

Sun or
Part Shade

Moist Well
Drained Soil

Frost Hardy

Papaver commutatum
'LADYBIRD'

Flowering June to August, this distinctive hardy annual poppy produces cup-shaped blooms with a big black blotch on hairy, green upright stems. They prefer full sun and a well-drained soil.

PLANT WITH

Papaver orientale
'Patty's Plum'

Team your bright-red ladybird poppies with this elegant pink-flowered perennial for an almighty colour pop from May to July.

H85cm x S40cm

Allium
'Globemaster'

The majestic spherical heads of starry purple flowers provide punctuation in a border in July.

H80cm x S20cm

Cotinus coggygria
'Royal Purple'

The dark purple foliage of this handsome deciduous shrub make a fine foil for red flowers, bearing its own pink, smoke-like blooms July to August.

H5m x S5m

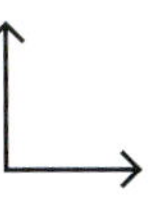

H45cm x S15cm

Sun

Moist Well Drained Soil

AGM Rated

Rosa

'DARCEY BUSSELL'

A compact, bushy, repeat-flowering shrub rose with a romantic blowsy look. Stems are smothered in blooms from June to August, the flowers aging to a darker burgundy hue. Plant in full sun or part shade and a moist, well-drained soil. Mulch once a year with organic matter in spring. Prune off spent flowers to encourage more.

PLANT WITH

Astrantia
'Claret'

Pretty, ruby-red pincushion flowers on tall, almost-black stems, flowering June to August.

H90cm x S30cm

Euphorbia characias
subsp. *wulfenii*

Vivacious evergreen shrub with billowing acid-yellow flowers March to July.

H1.2m x S1.2m

Calamagrostis x acutiflora
'Karl Foerster'

Stiff, upright grass with feathery flowers June to August. Excellent partner for a feathery screen, fading to buff.

H1.8m x S60cm

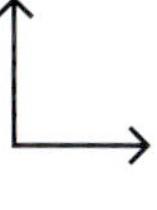

H1.2m x S90cm

Sun or Part Shade

Moist Well Drained Soil

Frost Hardy

Helenium

'MOERHEIM BEAUTY'

The distinctive daisy-type flowers of this orange-red perennial have a Mexican-hat look about them, with a central nose that's dusted in pollen – hence the common name, sneezeweed. Flowers are held on tall, wiry stems from June to August, performing best in fertile, well-drained soil in full sun. Deadhead to a branching side shoot and cut back once flowering has finished to prevent mildew.

PLANT WITH

Persicaria amplexicaulis
'Dikke Floskes'

Easy, reliable, clump-forming perennial with red drumstick flowers July to October.

H1.2m x S1.05m

Salvia nemorosa
'Caradonna'

Dark-purple spires from June to October, ideal for any sunny spot.

H50cm x S30cm

Heuchera
'Forever Red'

Intensely coloured ground cover plant with ruffled evergreen foliage. Small white flowers June to August. A versatile partner for any red or pink-flowered perennial; tolerant of sun or part shade and a moist but well-drained soil.

H40cm x S40cm

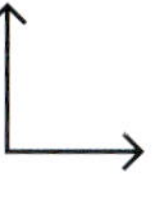

H1.25m x
S60cm

Sun

Moist Well
Drained Soil

Frost Hardy

AGM Rated

Astrantia major

'CLARET'

Dark ruby-red, pin-cushion flowers June to August on tall, wiry stems with a ruff of bracts (modified leaves) around the flower head. Water well in dry weather and mulch with organic matter to conserve moisture in spring. Once established, can be cut right back after flowering to produce another flush of flowers. Plant in rich, moist soil in part shade.

H90cm x S30cm

Sun or Part Shade

Moist Well Drained Soil

Pollinators

Frost Hardy

AGM Rated

PLANT WITH

Rosa
'Darcey Bussell'

Compact, bushy, repeat-flowering shrub rose with a romantic blowsy look.

H1.2m x S90cm

Geranium
'Rozanne'

Reliable, easy, clambering cranesbill with purple-blue flowers with a white centre.

H60cm x S80cm

Imperata cylindrica
'Rubra'

Dramatic Japanese blood grass that grows best in full sun and well-drained soil. Not evergreen. Flowers June to August.

H50cm x S30cm

Monarda

'CAMBRIDGE SCARLET'

Also known as 'bee balm' or 'bergamot', monardas' intricate flower heads comprise a number of curving flower tubes growing out of the centre point, creating a shaggy look from June to August. Bees and butterflies love these nectar-rich blooms. Grow in moist but well-drained soil in sun or part shade and mulch annually with organic matter.

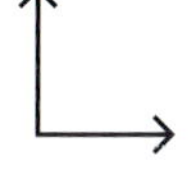

H90cm x S45cm

Sun or Part Shade

Moist Well Drained Soil

Pollinators

Frost Hardy

PLANT WITH

Lobelia cardinalis

Perennial bearing showy red flowers June to October – moisture lover so ideal for damp soil.

H90cm x S45cm

Rudbeckia fulgida var. *sullivantii* 'Goldsturm'

Dazzling golden, daisy-like flowers with cone-shaped noses, August to October.

H60cm x S45cm

Physocarpus opulifolius 'Diabolo'

Colourful deciduous shrub with red to purple foliage and clusters of pink-flushed white flowers June to August, followed by glossy red seedheads. Needs sun or part shade in moist but well-drained, slightly acidic soil.

H2m x S2.5m

Salvia greggii
'LIPSTICK'

An upright shrubby perennial with deep pinkish-red flowers and aromatic leaves that have a minty smell when crushed. Flowers emerge June to October, provided you deadhead regularly. These sun-loving plants are drought-tolerant once established, but semi-evergreen so they may lose their winter leaves in a cold garden.

PLANT WITH

Cosmos atrosanguineus

Fragrant chocolate cosmos bears velvety dark-red flowers on tall, wiry stems June to October.

H60cm x S45cm

Dahlia
'David Howard'

Popular decorative orange dahlia with dramatic dark leaves ideal for a hot border, July to October in sun.

H75cm x S75cm

Phormium
'Maori Queen'

Stripy, strappy, sword-like evergreen with bronze-green foliage and a rose-red margin. Architectural beauty for a sunny spot.

H1m x S1.5m

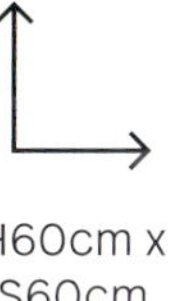

H60cm x
S60cm

Sun

Moist Well
Drained Soil

Pollinators

Frost Hardy

Crocosmia

'LUCIFER'

This feisty perennial emerges as ribbed, sword-like strappy foliage from a corm each spring, eventually producing an exotic flower (August to September) that resembles a brilliant red bird in flight. The knobbly, traffic-light seedheads, green to rusty orange, are just as exciting, and long-lasting too. The elegant foliage has a tendency to lean in clay soil – so plant the corms deep and in large swathes among upright shrubs or with discrete plant supports. They are happy in full sun or shady conditions, in moist but well-drained soil.

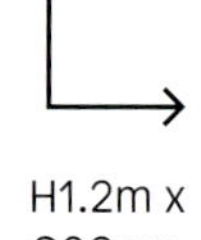

H1.2m x S80cm

Sun or Part Shade

Moist Well Drained Soil

Frost Hardy

AGM Rated

PLANT WITH

Amaranthus caudatus

Tassel-like, dangly flowers hang on this distinctive 'love-lies-bleeding' annual plant, flowering June to September.

H1.5m x S75cm

Crocosmia
'Emily McKenzie'

Orange-flowered 'Lucifer' relative, blooming July to September. Great for bold drifts and exotic plantings.

H60cm x S80cm

Ricinus communis
'Carmencita'

Bruised-looking, purple-bronze palmate foliage gives this exotic plant a real jungly look, with showy red flower spikes July to September.

H1.2m x S90cm

Sanguisorba officinalis

'TANNA'

With its button-like, bobbly deep maroon flowers on wiry stems, held above clumps of lacy green foliage, this eye-catching perennial flowers June to September in sun or part shade. Also known as 'great burnet', it's a robust ground-cover plant that performs best in moist, well-drained soil but is tolerant of dry soil too.

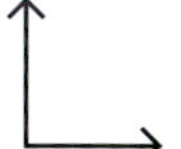

H50cm x S40cm

Sun or Part Shade

Moist Well Drained Soil

Pollinators

PLANT WITH

Dahlia
'Kilburn Glow'

A decadent waterlily type in magenta-rose, flowering June to the frosts (October). Prefers moist but well-drained soil and full sun.

H1.2m x S60cm

Achillea
'Cloth of Gold'

Bright-yellow, flat-headed umbel flowers on tall stems. Long-lasting and drought resistant.

H1.2m x S45cm

Hakonechloa macra

Ornamental Japanese grass that forms shaggy-looking cascades at the front of a border. Cultivar 'Nicolas' blushes red in autumn.

H35cm x S40cm

Hemerocallis
'STAFFORD'

This charming daylily bears masses of starry red, lily-like flowers with a central gold stripe leading into the flower throat. Flowering begins in July and continues through mid-summer; keep deadheading to prolong flowering. Foliage forms a clump of bright-green strappy foliage that's evergreen in mild areas. Plant in full sun in a moist, well-drained soil. Mulch with organic matter and water in spring to encourage buds to form.

PLANT WITH

Achillea
'Terracotta'

Sun-loving, colour-shifting perennial with flat umbel flowers that start terracotta but fade to creamy yellow.

H1.1m x S40cm

Helenium
'Waltraut'

Orange-yellow, daisy-type flowers on upright stems July to September.

H1m x S60cm

Nandina domestica

Heavenly bamboo is a compact, rounded evergreen shrub with elegant pointy leaves that give it a Japanese look. Cultivars such as 'Fire Power' and 'Blush Pink' will turn flame-red in autumn.

H45cm x S60cm

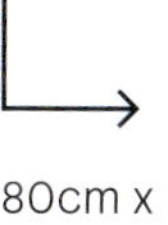

H80cm x
S45cm

Sun

Moist Well
Drained Soil

Frost Hardy

AGM Rated

Dahlia

'BISHOP OF LLANDAFF'

Bold but simple semi-double-flowered dahlia punching well above its weight in a hot border scheme, flowering July to October. Deeply divided bronze-red foliage. Ideal for a sunny site in humus-rich fertile soil. After the first autumn frost, it's best to lift the tubers with a garden fork and overwinter in a frost-free place to avoid them rotting in cold, wet soil.

PLANT WITH

Salvia coccinea
'Hummingbird'

A compact tender annual with eye-popping scarlet flowers. A handsome bedding partner for red dahlias.

H75cm x S30cm

Agapanthus africanus

Striking blue spheres of trumpet-shaped florets on stout stems among strappy green foliage. African lilies bloom June to September.

H60cm x S45cm

Ricinus communis
'Carmencita'

Dazzling castor oil plant with bold palmate leaves in reddish-purple, and bright-red sputnik-shaped fruits containing toxic seeds. Flowers July to September.

H1.2m x S90cm

H1.1m x
S45cm

Sun

Moist Well
Drained Soil

Pollinators

AGM Rated

Clematis

'REBECCA'

Large-flowered red clematis in pruning group 2 (cut back in late winter/early spring to healthy-looking buds) with a central boss of creamy yellow whiskers. Flowers emerge June to September, aging from plum to bright red – plant in full sun to light shade to get the best hue. Foliage is deciduous.

PLANT WITH

Fuchsia magellanica

Hardy fuchsia with red flowers and purple underskirts. Best in fertile moist but well-drained soil, flowering July to October.

H2.5m x S2m

Clematis
'Etoile Violette'

Vigorous purple clematis (pruning group 3), flowering July to September. Will scramble through trees or shrubs.

H5m x S1.5m

Hedera helix
'White Ripple'

Attractively variegated ivy with a white margin, ideal for a tricky north-facing wall. Best in light shade and appealing to wildlife.

H3m x S1m

H2.5m
x S1m

Sun or
Part Shade

Moist Well
Drained Soil

Frost Hardy

Cotoneaster horizontalis

Prized for its herringbone lattice of stems, this useful spreading, deciduous shrub has neat pinkish-white flowers May to June, then produces attractive red berries from September to November. It performs well in part shade but prefers full sun and can be trained up a sunny wall. It's bird and pollinator friendly, but considered an invasive non-native plant, so be careful to dispose of prunings carefully.

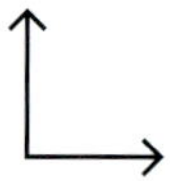

H1m x S1.5m

Sun

Moist Well Drained Soil

Pollinators

PLANT WITH

Clematis
'Rebecca'

Plum-red flowered clematis, aging to bright red from June to September. Plant in full sun to light shade to get best hue.

H2.5m x S1m

Acer palmatum
'Bloodgood'

Japanese maple producing red-purple foliage, turning redder in autumn. Full sun or part shade.

H4m x S4m

Euonymus europeaus
'Red Cascade'

Deciduous hedgerow shrub that makes a fine focal point in autumn, with its dainty orange and reddish-pink winged fruit.

H3m x S2.5m

Euonymus europaeus

'RED CASCADE'

This large deciduous hedgerow shrub was once harvested for making spindles – hence the common name of spindle. Small summer flowers from May to June are followed by dainty orange and reddish-pink winged fruit that last well into autumn. In winter the leaves turn bright red. Plant one in full sun for best colour, or part shade, in moist well-drained soil.

H3m x S2.5m

Sun or Part Shade

Moist Well Drained Soil

Pollinators

AGM Rated

PLANT WITH

Ricinus communis
'Carmencita'

Bold palmate leaves in reddish-purple, and bright-red sputnik-shaped fruits containing toxic seeds. Flowers July to September.

H1.2m x S90cm

Cornus alba
'Sibirica'

Distinctive dogwood with a framework of upright, fiery red stems after leaf-fall. Prune hard in late winter to promote vigorous red stems the following year.

H2.5m x S2.5m

Hakonechloa macra

Shaggy-looking ornamental grass. Cultivar 'Nicolas' blushes red in autumn; 'Aureola' is more gold.

H35cm x S40cm

Skimmia japonica

'RUBELLA'

A compact evergreen (male) shrub with handsome dark-red flower buds November to March, which open to white flowers April to May. Young plants are ideal as the low-maintenance anchor in winter containers, preferring moist but well-drained soil in part to full shade. Trim plants after flowering and mulch annually with organic matter.

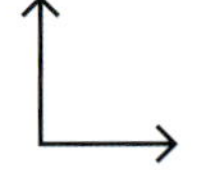

H1.5m x S1.5m

Part Shade or Full Shade

Moist Well Drained Soil

Pollinators

AGM Rated

PLANT WITH

Calluna vulgaris
'Garden Girls'

Upright spikes of deep-pink flowers July to November, and evergreen ground cover with plenty of texture over winter. Red and white cultivars are also available in this series. Prefers acidic soil or ericaceous compost.

H50cm x S60cm

Galanthus nivalis

Dainty nodding white winter flowers January to March, ideal for naturalizing in a sunny spot under deciduous trees and shrubs.

H10cm x S10cm

Photinia x fraseri
'Red Robin'

Glossy evergreen bearing vibrant red young leaves. For best colour, plant in full sun, but tolerant of part shade. Ivory flowers April to May.

H4m x S4m

Hamamelis x intermedia

'DIANE'

A reliable and sweetly fragrant late-winter-flowering shrub or small tree bearing red, spidery flowers on bare branches in January to February. Plant in well-drained, slightly acidic soil in full sun to part shade. Plants are deciduous and provide a good autumn show, the leaves turning from green through yellow, orange and red.

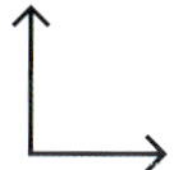

H4m x S4m

Sun or Part Shade

Moist Well Drained Soil

Pollinators

AGM Rated

PLANT WITH

Viola cornuta
'Red Blotch'

This eye-catching perennial pansy will shrug off winter weather, flowering September right across winter into May. Plant in a cool, part-shady spot. The petals are edible, too.

H15cm x S10cm

Cylamen coum

Available in white or pink forms, these tubby-flowered little gems have attractive silvery foliage. Plant in sun or part shade for flowers January to March.

H10cm x S10cm

Euphorbia characias subsp. *wulfenii*

Vivacious evergreen shrub flowering March to July. In winter the geometrically neat foliage comes into its own.

H1.2m x S1.2m

Build Your
Plant Matrix

+
+
+
+
+
+
+
+

Red with...

Red

Crocosmia ‘Lucifer’ with *Lobelia cardinalis*

Pink

Sanguisorba officinalis ‘Tanna’ with *Cosmos bipinnatus* (see page 168)

Purple

Red poppy ‘Ladybird’ and *Geranium* ‘Rozanne’ (see page 110)

Blue

Tulipa ‘Doll’s Minuet’ with *Myosotis sylvatica* (see page 18)

Green

Rosa ‘Darcey Bussell’ with foliage of *Bergenia cordifolia* (see page 48)

Yellow

Cornus alba ‘Sibirica’ with *Narcissus* ‘Tête-à-Tête’ (see page 108)

Orange

Dahlia ‘Bishop of Llandaff’ or ‘Spartacus’ with *Tithonia* ‘Torch’ (see page 238)

White

Salvia greggii ‘Lipstick’ with *Aquilegia vulgaris* var. *stellata* ‘Green Apples’ (see page 156)

Pink

PINK is the colour of joy and playfulness, party celebrations and sugary treats with sweet, fruity flavours. It is strawberry bonbons, flamingoes at sunset, iced doughnuts and raspberry ripple ice cream. Above all, it is Fun.

It is also the traditional colour of femininity and, like it or not, pink is used remorselessly to market toys to little girls – and their mothers. Perhaps the most famous pink branding belongs to teen doll Barbie, launched by Mattel in 1959. Barbie's iconic Dream House is decorated and furnished almost entirely in pink – as is the human-size Airbnb version in Malibu, California. While many of us would consider this a little too much, from a marketing point of view at least, Barbie Pink (Pantone 219) is the colour of global corporate success.

In the garden, pink is probably the easiest of colours to shop for and accommodate. There are so many pink-flowered plants to choose from that it can be hard to edit your choices and there's always the temptation to squeeze more in. Often, less is more in a small garden.

To organize your pink-themed border, avoid partnering the blue-pinks with warmer coral and salmon-pinks, unless you're using lots of neutral green foliage in between. One clever idea is to pair up light and dark pinks to create a more three-dimensional effect – teaming palest baby-knitwear pink with a more vivacious, adult 'lipstick' hue. Blush and Bashful if you will. They go surprisingly well together, forming the archetypal harmonious colour scheme.

There is strong purple-pink too – magenta. It's hard to believe that the colour was once denounced as 'malignant magenta' by the famous Victorian plantswoman Gertrude Jekyll, largely because as a synthetic textile dye, the colour relied on toxic chemicals in its manufacture. Thus, it represented to Gertrude and her friends in the Arts & Crafts Movement all that was wrong with unchecked industrialization at the time. Happily, today the hue has been redeemed for widespread garden use – black-eyed, magenta-flowered *Geranium* 'Ann Folkard' is one of the prettiest examples, endorsed with an Award of Garden Merit (AGM) from the RHS.

There can be few brighter, more cheerful ways to kickstart the year than by planting strident pink tulips under the blossoming canopy of a magnolia, cherry or apple tree. Pink blossom looks utterly lovely against a clear blue spring sky, and there are several smaller fruit trees that will fit the bill where garden space is limited. Cultivars such as *Prunus* 'Pink Shell' (H4m) or 'Pink Perfection' (H5m) flower from April, while crab apple *Malus* 'Candymint' (H5m) has red buds that open to blush-pink in May.

Alongside this fluttering confetti are pink tulips such as vivacious

'China Pink', darker 'Merlot' and delicate 'Angelique'. Plant them in autumn (November is best) in sunlit drifts or colour blocks, according to preference. You can hide their sometimes-tatty leaves by planting them among early spring ground cover, such as aquilegias, wallflowers (*Erysimum*), honesty (*Lunaria*), forget-me-nots (*Myosotis* – available in blue, white or pink) and *Brunnera macrophylla*.

In early summer, pink-flowered roses and clematis can create a picturesque cottage-garden look, for instance by training them on trellis around a rustic front door, or tumbling over a white-painted picket fence. You can also transform a part-shady spot with pink foxgloves, evoking a dappled woodland glade.

By midsummer, another particularly effective idea is to layer pink and white flowers with other pastel-toned blooms in palest purple and coolest sky blue. Think of pink and white foxgloves with pops of blue from campanulas and *Linum perenne*, iris 'Jane Phillips' and pale pink roses, aquilegia 'Nora Barlow' and *Persicaria bistorta* 'Superba'. This gentle, 'sorbet-style' of planting was demonstrated by Hay-Joung Hwang in her 2016 Chelsea Flower Show garden – a memorable confection so soothing and delicate it was almost dreamlike.

As if there weren't enough pink-flowered perennials for summer, there are plenty of exquisite annuals you can grow from seed on a windowsill too – such as *Cosmos bipinnatus* with its open flowers and fern-like foliage; pink *Cleome*, with its spidery 'firework' blooms; and the ruffled can-can flowers of sweet peas. Contrast them with rotund pon-pon dahlia blooms or pink spires of *Lythrum salicaria*.

Late summer also sees the arrival of *Echinacea purpurea* and the asters – or *Symphyotrichum* as many of them are now called. Good pink cultivars include pastel pink 'Patricia Ballard' and lipstick-pink 'Royal Ruby'; both will last well in a mild winter.

But when the frost really starts to bite, you can't beat winter-flowering shrubs such as deliciously fragrant *Viburnum bodnantense* 'Dawn' and *Daphne odora* 'Aureomarginata'. 'Dawn' holds her flowers in fragrant clusters on bare stems from November to March, while 'Aureomarginata' is about half the size, flowering from January to March, with dark, evergreen, gold-margined leaves. In either case, their fragrance is most effective in a sheltered sunny spot, close to a path or by the front door where wafts of scent will tantalize every visitor.

Bergenia cordifolia
'PURPUREA'

Flappy, thick-leaved elephant's ears are prized for their easy, shade-tolerant evergreen groundcover and cerise-pink spring flowers, March to April. Cultivar 'Purpurea' has glossy green leaves that turn reddish pink in winter, adding to the seasonal display. Growing from knobbly ground-level rhizomes, it's tolerant of sun or light shade and poorer soils.

PLANT WITH

Cyclamen coum

Attractive little pink flowers with petals like swept-back butterfly wings, blooming in the dead of winter, December to March, with attractive silvery-green foliage.

H10cm x S10cm

Helleborus niger

White hellebores are a sophisticated option for any shady border. The 'Christmas rose' blooms January to February, so will provide an elegant forerunner for the pink *Bergenia* blooms from March to April. The foliage partnership will look fab all year.

H30cm x S45cm

Berberis thunbergii 'Atropurpurea'

A handsome deciduous foliage plant for a well-drained, part-shaded spot, turning redder in autumn. Pale yellow flowers in mid-spring.

H1m x S2.5m

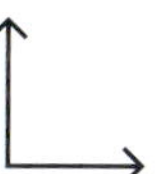

H60cm x
S75cm

Sun or
Part Shade

Moist Well
Drained Soil

Frost Hardy

Ribes x beatonii

This dainty flowering currant bears delicate trusses of soft yellow and pinkish-red flowers March to April – much loved by early pollinators and providing an eye-catching focal point in a sunny garden corner. *Ribes sanguineum* is another good choice, with lovely cerise pink flowers. Plant in full sun in a moist well-drained soil.

PLANT WITH

Tiarella 'Spring Symphony'

Foam flower bears lots of starry little flower spikes with a dark splotch in the centre of its deeply lobed leaves. Flowers May to July.

H25cm x S25cm

Lamprocapnos spectabilis

Available in white or pink forms, some with golden leaves, this enchanting perennial blooms April to May – its dangly flowers echoing the *Ribes*.

H1.2m x S50cm

Buxus sempervirens

Three neat little box balls (or one of their robust lookalikes, such as *Ilex crenata* or *Euonymus* 'Jean Hugues') will provide a smart contrast of form when grown under the spreading canopy of any *Ribes*. Lovely backdrop for *Tiarella* too.

Keep clipped to around H50cm x S50cm

H2.5m x
S2.5m

Sun

Moist Well
Drained Soil

Pollinators

Frost Hardy

Lamprocapnos spectabilis

More commonly known as dicentra, these dangly love-heart flowers are unbelievably pretty and intricate, emerging on arching stems from April to May. The foliage is deeply cut and fern-like, adding to their appeal, but the plant does die back after flowering. Best planted at the front of a border or en masse – those little flowers can get lost in a bigger planting scheme. Also available in white.

PLANT WITH

Saxifraga urbium

The tiny white blooms of London Pride, May to July, will create an attractive froth in sun or shade, floating on branching stems above a rosette of evergreen leaves.

H30cm x S70cm

Euphorbia epithymoides

Cushion spurge carries clouds of bright green to acid-yellow flowers in a neat mound. Flowering April to May, its starry blooms will appear alongside the hearts of *Lamprocapnos*.

H40cm x S60cm

Bergenia cordifolia 'Purpurea'

Producing upright pink flowers from March to April, the evergreen foliage will serve to flesh out this neat planting matrix, blushing purple in winter.

H60cm x S75cm

H1.2m x S50cm

Sun or Part Shade

Moist Well Drained Soil

Frost Hardy

AGM Rated

Magnolia x loebneri

'LEONARD MESSEL'

An elegant addition to any spring border, this magnolia cultivar has dark pink buds that open to blush-pink flowers in April. It's always an 'event' when magnolia blooms emerge – these ones start as upright cups then splay into stars. The open flower structure is said to make the blooms less susceptible to frost damage. Tolerant of chalky soils.

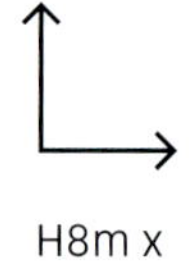

H8m x S6m

Sun or Part Shade

Moist Well Drained Soil

Frost Hardy

AGM Rated

PLANT WITH

Tulipa
'China Pink'

Dashing, lily-flowered tulip whose goblet-shaped blooms will echo those of your pink magnolia tree, but flowering slightly later (May). A sturdy cultivar loved by designers.

H50cm x S10cm

Magnolia stellata

If you have space for one magnolia, perhaps you have space for two? This classic white magnolia is a neat, popular option thanks to these stunning white, starry flowers, March to April.

H3m x S4m

Dryopteris affinis

Ferns will flesh out any part-shady planting matrix. This semi-evergreen, statuesque option has golden-brown scales and bright green young fronds that unfurl from the crown in spring.

H1.5m x S1m

Tulipa

'CHINA PINK'

These lily-shaped pink flowers are a garden-designer favourite, teaming well with darker tulips such as almost-black 'Queen of Night' or pink Triumph type, 'Don Quichotte'. Plant in late autumn in a sunny spot, about 15 to 20cm deep in moist but well-drained soil. Flowers appear in May and are relatively sturdy and upright.

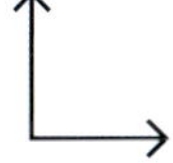

H50cm x S10cm

Sun

Moist Well Drained Soil

Frost Hardy

AGM Rated

PLANT WITH

Tulipa
'Don Quichotte'

Create a blush-and-bashful colour combination with 'China Pink' by planting them with these slightly darker pink 'Don Quichotte' tulips (April to May). Plant them deep and protect with an upturned wire hanging basket if you have squirrels in the garden.

H55xm x S10cm

Myosotis sylvatica

Forget-me-nots are biennial, and rampant self-sowers, but lovely nonetheless – offering clumps of foliage to hide tatty tulip leaves and a haze of blue flowers April to June.

H30cm x S15cm

Euphorbia palustris

The acid-green flowers and foliage of this architectural spurge make it a useful partner in sun or part shade. A herbaceous perennial that will die back in winter.

H1m x S1m

Clematis

'ROSAMUNDE'

There are lots of pink clematis to choose from; this pretty option has pink-on-pink striped flowers June to September. Belonging to pruning group 3, it flowers on the current year's growth, so needs to be cut back in late winter or early spring to keep flowers at eye level. Ideal for growing up a white, wrought-iron obelisk for some old-world cottage style, or on trellis to hide a boring fence panel. For dappled shade or full sun in a moist well-drained soil.

PLANT WITH

Aquilegia
'Nora Barlow'

Granny's bonnets are exquisite flowers for a cottage-style scheme. This one has fully double flowers that fade from deep pink toward white tips. Blooms are at their best in June.

H90cm x S40cm

Rosa
'New Dawn'

Shown here with *Clematis* 'Perle d'Azur', pale-pink rose 'New Dawn' is a popular climber that flowers July to September, in sun or part shade.

H3.5m x S2.5m.

Hosta sieboldiana var. *elegans*

The ribbed, slightly crimped-looking sculptural foliage of this impressive hosta is an attractive blue-green hue. Flowers are pale mauve July to July.

H1m x S1.2m

H2m x
S1m

Sun or
Part Shade

Moist Well
Drained Soil

Pollinators

Frost Hardy

Digitalis purpurea

Our native foxglove is such a reliable, shade-tolerant garden plant everybody should grow them! Although biennial (producing just a rosette of leaves in their first year, flowering the second) you can easily get these self-sowers to establish by shaking the spent flower heads in a part-shady spot at the end of year two. The tall spires add vertical interest, bearing flowers June to July – often with a bee's furry bottom poking out.

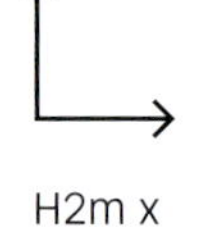

H2m x S60cm

Sun or Part Shade

Moist Well Drained Soil

Pollinators

Frost Hardy

PLANT WITH

Aquilegia 'Nora Barlow'

An attractive cottage-garden partner for foxgloves. Fully double flowers mean they're less interesting to bees, but beautiful nonetheless.

H90cm x S40cm

Digitalis purpurea f. *albiflora*

What could be better than partnering your pink foxglove with white foxgloves – the more the merrier to create a dynamic woodland edge effect in dappled shade, June to July.

H1.8m x S60cm

Polystichum setiferum

This shade-loving evergreen soft shield fern brings a relaxed and verdant look to any woodland-type planting scheme. Cut back tatty leaves at the start of the year to make way for new growth.

H1.2m x S90cm

Astrantia major

'ROMA'

This lovely clump-forming perennial is tolerant of part shade and clay, making it a useful addition in these tricky conditions, flowering June to September. It looks good with grasses and other cottage-garden perennials, with its pin-cushion flowers on wiry stems. It's said to be Dutch plantsman Piet Oudolf's favourite for his dazzling prairie planting schemes.

H60cm x S40cm

PLANT WITH

Sanguisorba officinalis
'Pink Tanna'

Cute, buzzy little bottlebrush flowers that bob about on wiry stems from June to September. Fern-like foliage.

H90cm x S50cm

Geranium
'Brookside'

Deep purple-blue flowers with maroon veins and a white centre, flowering May to August. Similar to 'Rozanne' but with smaller leaves.

H60cm x S45cm

Brunnera macrophylla
'Jack Frost'

Exciting, frosted-looking foliage plant that goes with everything. Delicate blue flowers April to May are a bonus – looking a bit like forget-me-nots. Also available with white flowers.

H40cm x S6cm

Sun or Part Shade

Moist Well Drained Soil

Frost Hardy

AGM Rated

Hydrangea macrophylla

'SWEET FANTASY'

Voluptuous mophead hydrangea, whose pinkish flowers are speckled dark pink, from June until the frosts. Grow in part or dappled shade and reliably moist but well-drained soil – don't let it go thirsty in drought. Good autumn colour – plants are deciduous – and best colour in alkaline soil.

PLANT WITH

Veronica spicata
'Rotfuchs'

With its German name translating to 'red fox', these bushy rose-pink spires lend a vertical note to billowing borders, June to August.

H40cm x S30cm

Eryngium giganteum

Prickly biennial sea holly with silvery thistle flowers bearing a ruff of spiky bracts and marbled lower foliage. Self-sows freely and flowers June to August. Prefers sun but will cope with part shade alongside the hydrangea.

H90cm x S30cm

Miscanthus sinensis
'Gracillimus'

Chinese silver grass whose leaf blades have a central silver stripe, and tassel-like flowers August to September.

H1.3m x S1.2m

H1.5m x
S1.5m

Part Shade

Moist Well
Drained Soil

Pollinators

Frost Hardy

Rosa

'GERTRUDE JEKYLL'

A hardy English shrub rose with plump pink buds unfurling into large, cup-shaped, fragrant double blooms that are packed with petals June to September. Blowsy and beautiful, it's named after a doyenne of Victorian horticulture, famous for her naturalistic colour schemes. Grow it in full sun for best results in a moist but well-drained soil. Roses love improved clay, so be sure to mulch and prune annually while dormant.

PLANT WITH

Lathyrus
'Prima Ballerina'

Fragrant annual sweetpea in pink and delicate purple June to August. Plant in moist, well-drained soil to clamber up a rustic obelisk, and deadhead to prolong flowering.

H1.8m x S30cm

Salvia nemorosa
'Caradonna'

Herbaceous perennial with rich purple flower spikes from June to October. Absolutely gorgeous alongside pink roses. Prefers full sun and a moist but well-drained soil.

H50cm x S30cm

Salvia nemorosa
'Silver carpet'

Distinctive silver foliage plant with velveteen leaves soft as lamb's ears. Drought tolerant and evergreen – perfect ground cover for sun or part shade. Can be vigorous in clay.

H20cm x S45cm

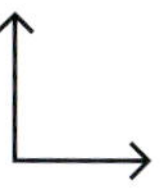

H1.5m x
S90cm

Sun

Moist Well
Drained Soil

Pollinators

Frost Hardy

AGM Rated

Phlox paniculata

'BRIGHT EYES'

Cottage-garden perennial with fragrant pink flowers that have a darker pink centre. Upright habit and long-lasting flowers from July to September. Loved by bees and other pollinators, in sun or part shade and moist, well-drained soil.

PLANT WITH

Geranium sanguineum
'Max Frei'

Bushy, pink-flowered geranium forming unrelenting mounds of flowers June to August. Plant in sun or part shade in any soil.

H45cm x S45cm

Echinacea purpurea
'White Swan'

Mix things up a bit by partnering your phlox with these drooping white daisies. Long-flowering from June to September in sun and moist, well-drained soil.

H60cm x S45cm

Stipa tenuissima

This good-natured wispy grass goes with absolutely everything. Blades turn from green to bleached blonde, with fluffy flowers June to October. It's semi-evergreen and you can thin out congested clumps by combing through with your fingers. A mindful activity – don't rush it.

H60cm x S30cm

H90cm x
S40cm

Sun or
Part Shade

Moist Well
Drained Soil

Pollinators

Frost Hardy

Echinacea purpurea
'MAGNUS'

These rosy-pink daisies emerge from June to September, with fuzzy, orange-brown nose cones that make them look a bit like a badminton shuttlecock. The flowers are held on strong, upright stems and are loved by butterflies – another excellent reason for growing them. Plant them in full sun for best results, in a moist, well-drained soil.

PLANT WITH

Persicaria amplexicaulis

Spikes of crimson-red flowers from July to October are held on wiry stems above semi-evergreen foliage. A peppy contrast for pink echinacea in sun or light shade and moist, well-drained soil.

H1.2m x S1.2m

Rudbeckia fulgida var. *deamii*

Sunny, golden black-eyed Susan with dark nose cone, flowering August to October with attractive seedheads to follow. Performs best in sun.

H60cm x S45cm

Miscanthus
'Flamingo'

Elegant grass with dark-pink tassel-like flowers in late summer. Best in a sunny, open spot.

H1.7m x S2m

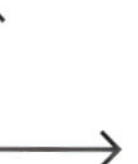

H1.5m x S45cm

Sun

Moist Well Drained Soil

Pollinators

Frost Hardy

Anemone hupehensis

'JAPONICA'

This charming, pink-flowered perennial arrives late, but is worth the wait! Flowering July to September, the tight, silky buds open to cup-shaped pink flowers on tall, wiry stems above lobed, semi-evergreen foliage. Plant in sun or shade in a moist, well-drained soil; cut back spent stems and tatty brown foliage after flowering. Other good Japanese anemones include 'September Charm' and dark-pink double 'Pamina'.

PLANT WITH

Colchicum autumnale
'Waterlily'

Autumn crocus with large, loose, goblet flowers in September and October. Available in single and double forms, including white. Ideal for naturalizing in a sunlit lawn.

H15cm x 10cm

Rudbeckia fulgida var. *sullivanti*
'Goldsturm'

Good-looking, late-season golden daisies, ideal for planting in swathes to contrast with your Japanese anemones. Flowers August to October in sun or light shade.

H60cm x S45cm

Calamagrostis x acutiflora
'Karl Foerster'

Upright feather reed grass ideal for informal screening, turning green to buff as the season progresses. Flowers June to August – best in full sun or light shade.

H1.8m x S60cm

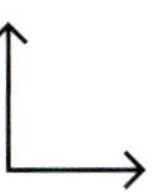

H1m x
S60cm

Sun or
Part Shade

Moist Well
Drained Soil

Pollinators

Frost Hardy

Symphyotrichum novi belgii

'ROYAL RUBY'

Rich, lipstick-pink aster producing compact clumps of foliage and cheerful daisy flowers from August to October. Plant in a well-drained soil in a sheltered, part-shady site for a glorious display well into autumn. Cut back hard after flowering and they'll rally for next year's display.

PLANT WITH

Echinacea purpurea

Eye-catching rose-pink daisies with petals that relax downwards and a central reddish-brown nose cone, flowering June to September. Plant in full sun in moist, well-drained soil.

H1.5m x S45cm

Aster x frikartii
'Mönch'

Charming lilac-blue flowered Michaelmas daisy with a sunny yellow eye. Flowers August to October and performs best in full sun.

H70cm x S40cm

Veronica
'Heartbreaker'

Pink-tinted hebes will add a distinctive architectural twist to a pink border, with their neatly aligned evergreen foliage and starry summer flowers June to August.

H60cm x S60cm

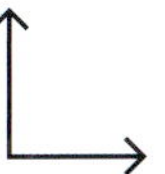

H50cm x
S50cm

Part Shade

Moist Well
Drained Soil

Pollinators

Frost Hardy

Hylotelephium spectabile
'BRILLIANT'

Vibrant, pink-flowered sedum that changes colour as the seasons progress, becoming darker pink towards autumn. Easy to grow and tolerant of shade and clay soils. Succulent green-grey leaves and tall, fleshy stems. To avoid drooping in late summer, grow them through an upturned wire hanging basket. Flowers August to November with good autumn structure. Good for pollinators – a real must-have all-rounder.

PLANT WITH

Eupatorium dubium
'Baby Joe'

Compact cultivar of towering *Eupatorium maculatum*. Joe pie weed has fluffy pink flowers that emerge August to September on upright, dense plants – the flowers are adored by pollinators of all kinds.

H70cm x S50cm

Euphorbia palustris

Bright-green flowers May to June and foliage that lasts well into autumn, turning attractive shades of yellow and orange. For an evergreen alternative, go for *Euphorbia amygdaloides* var. *robbiae*.

H1m x S1m

Pennisteum advena
'Fireworks'

Borderline-hardy, show-stopping grass that forms delightful fountains of pink foliage and fluffy flowers July to September. Best in a sheltered sunny spot with protection over winter.

H90cm x S60cm

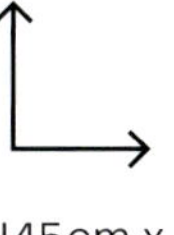

H45cm x
S45cm

Sun or
Part Shade

Moist Well
Drained Soil

Pollinators

Frost Hardy

AGM Rated

Cyclamen coum

These delicate-looking cerise-pink flowers can really draw the eye in a winter border with their swept-back petals and silvery, mottled leaves. They might be short in stature, and sweet, but they naturalize well and shrug off harsh winter weather. Plant them in part shade under deciduous shrubs and trees in humus-rich soil.

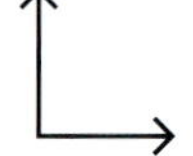

H10cm x S10cm

Part Shade

Moist Well Drained Soil

Frost Hardy

AGM Rated

PLANT WITH

Erica carnea
'Springwood Pink'

Vigorous dwarf evergreen shrub with dainty, bell-like flowers and a trailing habit, flowering January to March. Performs best in a sandy, well-drained soil that's slightly acidic.

H15cm x S45cm

Galanthus nivalis

Common native snowdrop with white nodding flowers and green markings January to March. Plant them 'in the green' if possible, just after flowering, rather than as dry bulbs. Lift and divide them to help your collection spread.

H10cm x S10cm

Heuchera
'Wild Rose'

Attractive berry-coloured plant whose semi-evergreen scalloped leaves have dark veins for contrast. Heucheras can provide a bright foil for cyclamen in winter – but opt for the brighter colours as the dark plum and black ones tend to recede against bare soil. Pink flowers June to August.

H30cm x S50cm

Daphne odora

'AUREOMARGINATA'

A good-looking evergreen shrub with clusters of fragrant pink flowers January to March. The glossy green leaves have a golden edge, adding valuable extra colour in moist, well-drained soil and a sunny sheltered or light shady position.

PLANT WITH

Erysimum
'Bowles's Mauve'

Bushy and vigorous, this pinky-mauve wallflower – a short-lived perennial – flowers February to July. Great value in any garden border, preferring full sun and a moist, well-drained soil.

H75cm x 60cm

Helleborus niger

The Christmas rose makes an enchanting partner for any winter flowering shrub. Trim away the foliage in January so you can see the flowers better, January to February. A good choice for shade, with leathery evergreen foliage.

H30cm x S45cm

Berberis thunbergii
f. atropurpurea
'Harlequin'

Deciduous shrub with reddish leaves marbled pink and white. Yellow spring flowers are followed by red berries later in the year.

H1.5m x S1.2m

H1.5m x
S1.5m

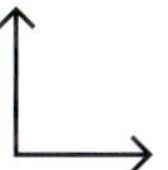

Sun or
Part Shade

Moist Well
Drained Soil

Frost Hardy

Build Your Plant Matrix

Pink with...

+

+

+

+

+

+

+

+

Colour	Combination
Red	*Tulipa* 'China Pink' with *Tulipa* 'Doll's Minuet' (see page 18)
Pink	*Bergenia cordifolia* 'Purpurea' with *Cyclamen coum*
Purple	*Echinacea purpurea* with *Verbena bonariensis* (see page 104) or *Salvia nemorosa* 'Caradonna' (see page 102)
Blue	*Rosa* 'Gertrude Jekyll' with *Eryngium x zabelii* 'Big Blue' (see page 138)
Green	*Lamprocapnos spectabilis* with *Euphorbia epithymoides* (see page 150)
Yellow	*Knautia macedonica* with *Achillea* 'Moonshine' (see page 102)
Orange	*Cosmos bipinnatus* with *Dahlia* 'David Howard' (see page 234)
White	*Digitalis purpurea* with *Digitalis purpurea* f. *albiflora* (see page 256)

Purple

PURPLE is the colour of magic and mystery, royalty and luxury. It is the colour of amethyst, summer-hazy lavender fields, thistles and blackcurrant juice. It hovers on the colour wheel between red and blue and is created by mixing those primaries together. Add a little too much red and you'll swing towards maroon and burgundy – warm, grapevine colours. Add a little too much blue and you create a cooler tone closer to mauve.

The truth is, these nuanced shades can look warm or cool depending on the sunlight (or in Britain, the lack of it), so in this chapter we'll focus on the broad array of both cool and warm purple flower tints.

Purple is a versatile partner in the garden: you can pair it with pink, blue or white without any real complaints, creating a harmonious arc on the colour wheel. Partner purple with crimson, however, and you have something altogether more exciting and daring. Think of those old-fashioned dangly fuchsia flowers – a red ballet skirt over a purple petticoat. Red poppies among purple geraniums will also dazzle. These colours together look extravagant, opulent and ecclesiastical.

Another bold 'colour pop' can be made by pairing purple with orange – agapanthus with orange crocosmia perhaps, or purple honesty with orange 'Ballerina' tulips. These complementary hues lie opposite each other on the colour wheel for maximum contrast, guaranteed to get the heart racing.

In spring, there are lots of exciting purples to get the season started – *Crocus tommasinianus* is a delightful little spring flower that can self-sow and naturalize over time into huge swathes in your lawn, the little stamens providing a spicy zing of yellow. Climbing wisteria is another winner for this time of year – a favourite for training up the front of a sunny house: plants in sunshine perform a lot better than those in shade. You can also plant them to clamber over a sunny pergola and create your own 'wisteria walk', dripping with purple pea-like flowers from May to June accompanied, if you like, by two other pretty purple stalwarts – allium 'Purple Sensation' and *Aquilegia vulgaris*.

There are lots of purple alliums to choose from – their spherical drumstick heads a mass of starry florets in early summer. There's *Allium giganteum*, which can reach giddy heights of 1.5 metres, and big-headed 'Globemaster', which can produce blooms the size of bowling balls. They make excellent partners for a wide range of May flowers, such as lupins and foxgloves, hence their popularity at RHS Chelsea Flower Show.

Once summer sets in, there are even more purple options. For sunny, drought-tolerant Mediterranean-style plantings, you can team lavender with purple nepeta, pale-purple bearded irises and geraniums, alongside silver-leaved eryngium thistles. The purple spires of *Salvia* x *sylvestris* 'Mainacht' (indigo-blue) or slightly later-flowering *Salvia nemorosa* 'Caradonna' (royal purple), will flower for ages in a sunny border – 'Caradonna' blooming right into October. For shady woodland gardens, go for lesser periwinkle (*Vinca minor*) – an evergreen ground cover plant that copes well in dry shade, and hardy geraniums such as 'Rozanne', which has flowers in purple-blue with a central white eye. In a cottage garden, purple roses, phlox and delphiniums will fit the bill perfectly.

As summer draws on it's the turn of asters – or *Symphyotrichum* as they're now called – to take centre stage. Purple ones include 'Purple Dome' and 'Violetta', and paler 'King George' and 'Mönch'. You might also like to use Russian sage *Perovskia* 'Blue Spire' and grasses among them. These good-natured plants will keep flowering alongside the likes of yellow rudbeckia long into autumn, provided frosts are light.

In autumn, it's the turn of the autumn crocuses, *Crocus autumnale*. These are also known by the common name of 'naked ladies' – a reference to the fact they'll often flower before being planted, straight from the bulb, in a rather indecorous fashion. They're a soft mauve-pink colour, adding a lighter note to an otherwise purple colour scheme.

To finish the year comes a host of cosy purples offered by the likes of winter pansies – such as *Viola wittrockiana* cultivars – and the delicate, pale-purple blue of sweet violet, *Viola odorata*, flowering alongside the dainty little flowers of early Algerian and reticulate irises, such as 'JS Dijt'.

Don't forget the many purple-leaved foliage plants and shrubs to use alongside your purple flowers. These range from the chocolate brown of *Pittosporum* 'Tom Thumb', *Physocarpus* 'Diabolo' and *Hylotelephium* 'Matrona' to the warmer burgundy red of *Cotinus coggygria* 'Royal Purple' and *Berberis thunbergii* f. *atropurpurea*. There's an excellent array of bruised-coloured heucheras to consider too: from 'Palace Purple' and 'Forever Purple' to 'Wildberry' and 'Plum Pudding'. Do remember that these dark-leaved plants can look quite recessive in a border and might not show up against the soil – for best results, plant them alongside green or gold, so they can sing.

Crocus tommasinianus

These open, goblet-shaped purple flowers have prominent yellow anthers (the bit that contains the pollen), pearl-white stems and grass-like foliage. They're lovely naturalized in lawn – especially when planted en masse in long swathes along a grassy path edge. Flowering February to March, they'll tolerate dappled shade under a light, shrubby canopy and the early bees will love them.

PLANT WITH

Viola odorata

These delicate little purple violets will in time form a semi-evergreen carpet of fragrant flowers blooming from March to April. Ideal for a woodland-style planting scheme.

H20cm x S30cm

Eranthis hyemalis

Golden winter aconites, with their choir-boy collar of green bracts, make a cheerful partner for purple tommy crocuses flowering January to February. Also good for naturalizing in part shade.

H8cm x S5cm

Polystichum setiferum

Evergreen soft shield fern that has dark green fronds that unfurl into a spreading array of tactile foliage. A shade lover for moist but well-drained soil. Cut back in early January.

H1.2m x S90cm

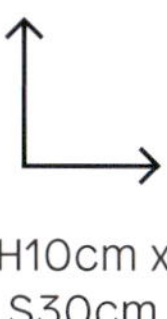

H10cm x
S30cm

Sun or
Part Shade

Moist Well
Drained Soil

Pollinators

Frost Hardy

AGM Rated

Viola odorata

Pretty, fragrant sweet violet is a semi-evergreen perennial that appreciates a sheltered spot. Ideal for covering the bare soil under deciduous shrubs and trees, its spring flowers in March and April are a welcome breath of fresh air. Plant in part shade in a moist, humus-rich soil.

PLANT WITH

Muscari armeniacum

Grape hyacinths have purple-blue flowers arranged like bobbly thimbles, with a paler tuft at the top and grass-like foliage. A good colonizer flowering in March.

H15cm x S10cm

Primula vulgaris

Our woodland primrose has pastel-yellow flowers March to May and looks stunning when planted en masse. Good for naturalizing in part shade.

H20cm x S35cm

Hedera helix 'Goldfinger'

This gold-variegated ivy is a bushy, shade-tolerant cultivar, with golden foliage whose colour brightens up with a bit of sun. Plant in moist, well-drained soil.

H1.5m x S1.5m

H20cm x S30cm

Part Shade

Moist Well Drained Soil

Frost Hardy

Fritillaria meleagris

These maroon-red snake's-head fritillaries have an exciting chessboard pattern on their nodding flowers, April to May. They're meadow wildflowers that prefer a reliably moist soil in sun or light shade – naturalizing and spreading year after year in the right conditions. Perfect for a woodland edge in the garden under the light canopy of deciduous shrubs or trees.

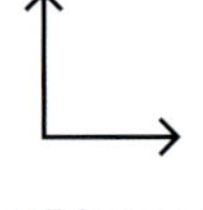

H30cm x S8cm

Sun or Part Shade

Moist Well Drained Soil

Frost Hardy

AGM Rated

PLANT WITH

Anemone blanda

Wood anemone bearing large daisy-like purple flowers March to April. Known as windflowers as they bob about in the winter breeze. Plant in sun or light shade.

H15cm x S15cm

Leucojum aestivum

Summer snowflake has a distinctly snowdrop look about it, but is much taller and flowers much later – in March to May. Plant in sun or light shade in moist, well-drained soil.

H50cm x S10cm

Hedera helix 'White Ripple'

A cool and compact white-and-green-variegated ivy that's loved by insects and the birds that feed on them. A versatile evergreen climber that's self-supporting in sun or shade. Nectar-rich flowers are followed by little berries loved by birds.

H3m x S1m

Aquilegia vulgaris

Charming perennial with nodding purple flowers, splayed sepals and hooked spurs, creating a demure, hooded look that hints at its common name: granny's bonnet. Plants are easy going and readily self-sow into paths and pavement cracks, flowering May to June. Plants hybridize freely so the offspring may not turn out purple. Plant in sun or part shade in moist well-drained soil.

PLANT WITH

Iris sibirica

Siberian flag iris whose purple-blue flowers have intricately veined yellow and white falls, and narrow, grass-like foliage. Prefers sun or part shade and moist, slightly acidic soil. Perfect in a pondside setting.

H1.2m x S50cm

Anthriscus sylvestris 'Ravenswing'

Planted en masse, our native cow parsley forms a froth of white umbels that billow in a breeze. The black-stemmed cultivar 'Ravenswing' is a more decorative option. Sun or part shade.

H1m x S30cm

Berberis thunbergii f. *atropurpurea* 'Harlequin'

Deciduous shrub with reddish leaves marbled pink and white. Yellow spring flowers are followed by red berries later in the year.

H1.5m x S1.2m

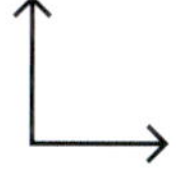

H1m x S45cm

Sun or Part Shade

Moist Well Drained Soil

Frost Hardy

Wisteria sinensis

This dramatic climber, when smothered in its cascading pea-like flowers from May to June, will always turn heads. The dangling racemes can transform a sunny wall – but they will need a wire support strung from secure vine eyes. It's a big plant if left to its own devices, so treat it to a twice-yearly pruning regime in late summer after flowering and midwinter to keep the whippy stems in check and stimulate the production of flower buds. Best in full sun and moist well-drained soil.

PLANT WITH

Allium
'Purple Sensation'

These upright, peppy purple spheres will provide percussive visual contrast for cascading wisteria flowers. Plant the bulbs in autumn in sun and a moist but well-drained soil.

H1m x S1m

Sambucus nigra
'Black Lace'

With its bold black dissected foliage and startling pink June flowers, this handsome elder will provide a striking backdrop ideal for a small garden. Plant the shrub in sun or part shade, in moist, well-drained soil.

H3m x S2m

Ophiopogon planiscapus
'Nigrescens'

Black mondo grass or lilyturf is a clump-forming evergreen – not technically a grass but with the same visual effect. With its purple flowers June to August, it's a dramatic partner to plant around the ankles of trees and shrubs, in sun or light shade.

H20cm x S30cm

H9m x
S5m

Sun

Moist Well
Drained Soil

Pollinators

Frost Hardy

AGM Rated

Lupinus
'MASTERPIECE'

Fabulous purple skyscraper flowers blooming from June above attractive star-shaped green foliage. A popular plant among show garden designers and ideal in a jewel-themed planting scheme. Prefers full sun and moist but well-drained soil.

PLANT WITH

Cirsium rivulare

Statuesque thistle flowering July to August, creating pops of bright cerise-pink that catch the eye when threaded through a purple or blue border. Best in full sun and moist, well-drained soil.

H1.2m x S60cm

Papaver orientale
'Patty's Plum'

Elegant oriental poppy flowering May to June, with plum-pink flowers and an architectural presence. Plant in sun and a well-drained soil.

H75cm x S45cm

Heuchera
'Wild Rose'

Sumptuous, purple-leaved foliage plant with scalloped leaves and dark maroon veining. Semi-evergreen ground cover for sun or light shade.

H30cm x S50cm

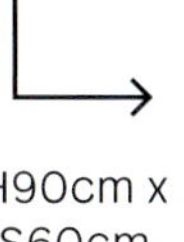

H90cm x S60cm

Sun

Moist Well Drained Soil

Pollinators

Frost Hardy

AGM Rated

Allium hollandicum

'PURPLE SENSATION'

No purple border is complete without the visual punctuation offered by these perennial bulbs. There are lots of alliums to choose from, but this one is a favourite thanks to its dense, reliably neat, starry flower spheres and intense colour in June. The grey-green foliage can become tatty, so team them with foliage plants that will hide their 'tankles' (tatty ankles).

PLANT WITH

Nepeta racemosa
'Walker's Low'

Fragrant, clump-forming perennial with purple flowers held on upright stems March to September. Plant in sun or light shade in moist, well-drained soil.

H60cm x S50cm

Lupinus
'Towering Inferno'

Fiery orange lupin forming plump spires June to July above clumps of attractive palmate foliage. Best in sunshine and a moist, well-drained soil.

H90cm x S60cm

Phormium
'Maori Queen'

This strappy-leaved evergreen has a dramatic, exotic look and vibrant foliage in bronze-green with a pink-red margin and cream edges. Best in a sunny sheltered spot.

H1m x S1.5m

H1m x
S1m

Sun or
Part Shade

Moist Well
Drained Soil

Pollinators

Frost Hardy

AGM Rated

Delphinium

'BLACK KNIGHT GROUP'

Regal-looking, purple-flowered perennial producing dark-eyed spires in June and July. These handsome plants are King Charles's favourites at his garden in Highgrove; they prefer a sunny position and look best when grown at the back of a deep border in moist but well-drained soil. Protect from snail damage when young and stake early.

PLANT WITH

Veronica longifolia 'Marietta'

Upright spikes in rich purple give this clump-forming perennial a voluptuous look, June to August. Best in sun and moist, well-drained soil.

H1m x S45cm

Alchemilla mollis

Lady's mantle has frothy yellow flower heads June to September that seem to float above neat green scalloped leaves, which hold water droplets like a bead of mercury. A delight on lots of levels and tolerant of drought too.

H50cm x S50cm

Stipa tenuissima

Wispy, clump-forming perennial grass that goes with everything. Flowers June to October maturing to blonde. Comb through in spring to remove dead stems.

H60cm x S30cm

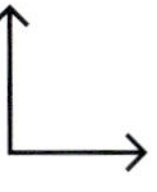

H1.7m x
S75cm

Sun

Moist Well
Drained Soil

Pollinators

Frost Hardy

AGM Rated

Lavandula angustifolia

'HIDCOTE'

Popular with pollinators, this compact English lavender carries fragrant purple flowers on silver-grey stems above aromatic foliage. Ideal for growing en masse as a hedge to line a path, its aroma will be released as you brush past. Keep the plants neat by trimming them back after flowering once the bees have lost interest – this will promote vigorous young growth the following year. Best in sun and a moist, well-drained soil.

PLANT WITH

Lavandula x *intermedia*
'Edelweiss'

Team your purple lavender with a white one such as Edelweiss. The foliage will sit well together and the contrasting flowers will add an interesting 'light and shade' effect, June to August.

H60cm x S60cm

Rosa
'Gertrude Jekyll'

A hardy English shrub rose with plump pink buds unfurling into large, fragrant bright-pink double blooms from June to the frosts.

H1.1m x S90cm.

Cotinus coggygria
'Royal Purple'

A large, burgundy-red-leaved deciduous shrub known as the smokebush thanks to its smoke-like pink flower plumes in July and August. Plant in sun and prune hard in March.

H5m x S5m

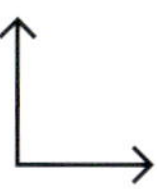

H60cm x S75cm

Sun

Moist Well Drained Soil

Pollinators

Frost Hardy

AGM Rated

Phlox paniculata

'BLUE PARADISE'

An attractive evergreen cottage-garden perennial flowering July to October. The purple-mauve flowers form fragrant domes on upright stems, attracting a wide range of opportunist pollinators. For best results, plant in sun or part shade in a moist but well-drained soil.

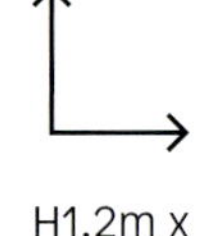

H1.2m x S1m

Sun or Part Shade

Moist Well Drained Soil

Pollinators

Frost Hardy

PLANT WITH

Agastache
'Blue Fortune'

Giant hyssop is loved by bees, producing vertical spires of purple flowers June to October. It's ideal in a sunny spot among Mediterranean or cottage-garden plants, or grasses.

H1m x S40cm

Monarda didyma
'Squaw'

Perennial bergamot that has distinctive pink flowers attractive to bees and butterflies. Its intricate flower tubes are arranged into a central whorl creating the look of an explosive mini firework. Sun or part shade.

H90cm x S45cm

Berberis thunbergii
f. *atropurpurea*

Reddish purple deciduous shrub with marbled leaves in burgundy and white. Makes a stunning hedge, becoming richer coloured in autumn.

H1m x S2.5m

Scabiosa

'BUTTERFLY BLUE'

These delicate-looking lilac-blue flowers are like mini pin cushions, held on wiry stems above the foliage, flowering July to September. The plants prefer a sunny spot in moist but well-drained soil and will draw in the pollinators. They have an enchanting meadow wildflower look but would be equally at home in gravel or a wildlife-friendly dry garden.

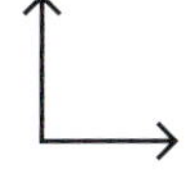

H40cm x S40cm

Sun

Moist Well Drained Soil

Pollinators

Frost Hardy

PLANT WITH

Campanula persicifolia

These lilac-blue bell flowers, held on tall stems, will harmonize with other soft purple flowers perfectly. They'll provide an attractive contrast of flower shape, from June to August in sun or light shade.

H80cm x S30cm

Verbascum 'Clementine'

An upright, tangerine-coloured mullein with pink stamens – a real showstopper flowering July to September. Sun or part shade.

H1.5m x 60cm

Festuca glauca 'Intense Blue'

Silvery blue tussock-forming fescue that needs full sun for best colour. Tolerant of drought once established, with steel blue flowers in June to July.

H30cm x S20cm

Clematis
'ETOILE VIOLETTE'

Regal, purple-flowering clematis belonging to pruning group 3, which means you can cut it back in early spring to about 15–20cm above ground level and it will bounce back to flower on the new stems that emerge. Don't forget to feed and mulch to help it recover. Flowering July to September, it's an attractive option for trellis or an obelisk, or to clamber through a large shrub or tree.

PLANT WITH

Lysimachia atropurpurea
'Beaujolais'

A short-lived perennial loosestrife with wine-coloured flowers and silver-green foliage. Attractive to bees and other pollinators and for sun or part shade.

H60cm x S50cm

Clematis
'Princess Diana'

A truly regal partner for 'Etoile Violette', grow this neon-pink, lily-flowered climber in sun or light shade and moist, well-drained soil. Flowers August to October.

H2.5m x S1m

Fagus sylvatica f. *purpurea*

Copper beech makes a stunning hedge when grown in staggered rows from bare-root plants. European beech *Fagus sylvatica* has marcescent leaves in winter (it retains the young foliage); this one has purple ones all year – best in sun.

Keep the hedge trimmed to less than H2m. Mature: H40m x S8m

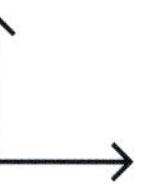

H5m x
S1.5m

Sun or
Part Shade

Moist Well
Drained Soil

Frost Hardy

AGM Rated

Echinops ritro
'VEITCH'S BLUE'

These spiky pompom flowers in metallic purple-blue are held above silver-green foliage and tall, rough stems. It's a striking architectural option for a purple border, and very attractive to pollinators. Best in well-drained soil in sun or part shade.

PLANT WITH

Eryngium bourgatii
'Picos Blue'

Contrast the globe thistles of echinops with these silvery sputniks, resembling NASA's International Space Station or something Darth Vader might launch from his Death Star. Plant in sun and moist, well-drained soil but also drought tolerant.

H50cm x S30cm

Rudbeckia fulgida var. *sullivantii*
'Goldsturm'

For contrast, these late-season, upright golden daisies bloom August to October. Grow in drifts in a sunny spot in moist but well-drained soil.

H60cm x S45cm

Physocarpus opulifolius
'Diabolo'

Dramatically named after the devil, this dark, bronze-leaved shrub makes a brooding backdrop for silver-purple flowers or lime-green ones.

H2m x S2.5m

H90cm x
S45cm

Sun or
Part Shade

Moist Well
Drained Soil

Pollinators

Frost Hardy

Salvia nemorosa
'CARADONNA'

Perennial sage, with its perky purple flower spikes from June to October, is a great-value plant much loved by garden designers. Producing fragrant mounds of grey-green foliage, you can cut the spent blooms and they'll produce another flush later in the season. A good-looking partner for a wide range of pink and blue-flowered plants as well as grasses. A must-have in sun and moist, well-drained soil.

PLANT WITH

Eryngium bourgatii
'Picos Blue'

Large-flowered, silvery sea holly on tall stems above attractive marbled foliage. Grow in sun and moist, well-drained soil.

H50cm x S30cm

Achillea
'Moonshine'

Flat umbels of bright yellow make a dazzling contrast for any purple planting scheme. Upright, with feathery foliage, best in sun and a moist, well-drained soil.

H60cm x S60cm

Cercis canadensis
'Forest Pansy'

Multi-stemmed deciduous tree with purple, heart-shaped leaves that turn yellow in autumn. Crimson flowers March to May.

H8m x S8m

H50cm x S30cm

Sun

Moist Well Drained Soil

Pollinators

Frost Hardy

AGM Rated

Agastache

'BLACKADDER'

Bee-friendly hyssop is ideal for creating a vertical accent in a cottage garden border, its tall, bottle-brush flowers emerging like purple rockets from July to October. Stems are slender and foliage aromatic. These are short-lived perennials that prefer a sunny position in a moist but well-drained soil.

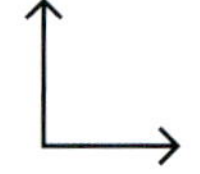

H90cm x S40cm

Sun

Moist Well Drained Soil

Pollinators

Frost Hardy

PLANT WITH

Verbena bonariensis

Easy-going, pink or light purple-flowered perennial whose tightly packed flower clusters are held on branching, wiry stems. Its sparse silhouette means you can plant it at the front of a border and still see what's behind. Beautiful and useful.

H2m x S45cm

Echinacea purpurea 'Magnus'

Pink daisy with lax petals and a prominent orange-brown nose cone, June to September. These upright flowers are loved by butterflies.

H1.5m x S45cm

Stipa tenuissima

This blonde grass billows in the breeze, bringing movement to any planting scheme – a good contrast for mounding perennials.

H60cm x S30cm

Perovskia

'BLUE SPIRE'

Russian sage is a tall, upright silvery sub-shrub with lavender-blue flowers July to October. Cultivar 'Blue Spire' is the popular (AGM) choice, flowering August to September, with aromatic grey-green leaves. It's a good choice for any purple-themed border or silver-foliage-led planting scheme.

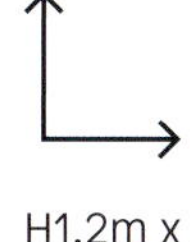

H1.2m x S1m

Sun or Part Shade

Moist Well Drained Soil

Pollinators

Frost Hardy

AGM Rated

PLANT WITH

Salvia nemorosa
'Caradonna'

Easy-going salvia with purple flower spikes blooming June to October. Best in light, well-drained soil and full sun.

H50cm x30cm

Echinacea purpurea
'Magnus'

Pink-flowered, upright daisy perfect for a naturalistic prairie-style planting scheme, flowering in late summer in sun and moist, well-drained soil.

H1.5m x S45cm

Panicum virgatum
'Shenandoah'

Red-tipped switch grass has upright foliage and masses of hazy pink flowers September to October. Lovely in a prairie scheme, offering a little height.

H90cm x S90cm

Colchicum autumnale

'WATERLILY'

Purple-flowered autumn crocuses are actually members of the lily family. They emerge from the soil without leaves, which is how they were given the titillating common name of 'naked ladies'. Best in light shade and a moist, well-drained soil, you can plant them in autumn (September to October) 8–10cm deep and leave them to naturalize.

PLANT WITH

Aster x frikartii
'Mönch'

Lavender-blue Michaelmas daisies smother this vigorous, mound-forming aster from August to October and last well to the frosts.

H70cm x S40cm

Nerine bowdenii
'Alba'

Starry, white-flowered South African plant, growing from bulbs and flowering September to November with strappy foliage.

H45cm xS30cm

Heuchera
'Plum Royale'

This dark-leaved coral bells is a pretty choice for purple planting schemes, with pink-white flowers July to August. Grow in part shade in reliably moist but well-drained soil.

H25cm x S40cm

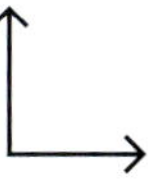

H15cm x
S10cm

Part Shade

Moist Well
Drained Soil

Pollinators

Frost Hardy

AGM Rated

Viola x wittrockiana

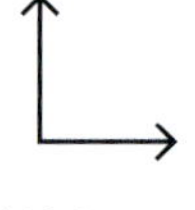

H00cm x S00cm

Sun or Part Shade

Moist Well Drained Soil

Pollinators

Frost Hardy

AGM Rated

Pansies are great for adding winter colour to pots and baskets, and this dark maroon cultivar is a little bit special, with its silky purple and white petals and yellow eye. Flowers keep going from October right on to May if you keep deadheading. Sun or light shade and a moist but well-drained soil, or peat-free compost if using a container.

PLANT WITH

Scilla siberica

Nodding lilac-blue spring flowers March to April, each with a blue stripe and strappy green foliage. Naturalize them in grass or under a deciduous shrub.

H20cm x S5cm

Narcissus 'Tête-à-Tête'

Dinky dwarf daffodil ideal for a naturalizing in a sunny outdoor spot, or for growing (or gifting) in a patio container, flowering March to April. H20cm x S5cm

Pittosporum tenuifolium 'Tom Thumb'

Purple-leaved, black-stemmed shrub with a compact, rounded habit. Needs shelter from the harshest winter weather and produces its best colour in sun.

H1m x S60cm

Iris reticulata

'J.S. DIJT'

Elaborate-looking reticulate irises hail from the mountainous regions of Turkey, Iran and Iraq, where they thrive in cool, sunny conditions in well-drained gritty soil. This maroon-flowered cultivar has a bright flash of yellow on its lower petals to catch the eye; they come in a range of cobalt and paler blues too. They need sun and well-drained soil, flowering January to February.

PLANT WITH

Viola odorata

Semi-evergreen sweet violet is a charming addition for your woodland area with fragrant flowers February to March. The flowers can be candied to decorate cakes or scattered on winter salads.

H20cm x S30cm

Galanthus nivalis

Snowdrops will make a dainty partner for reticulate irises, flowering January to March and offering a contrast of flower form.

H10cm x S10cm

Bergenia cordifolia
'Purpurea'

The fleshy, flappy-looking purple foliage of this bergenia cultivar will provide an excellent partner for maroon reticulate irises in winter. Heucheras too.

H60cm x S75cm

H12cm x S8cm

Sun

Moist Well Drained Soil

Frost Hardy

Geranium

'ROZANNE'

A popular hardy geranium that's a strong blue-purple and an example of a plant that could harmonize with blue or purple flowers, as you fancy. This long-flowering, cottage-garden perennial blooms June to October; plants are lax and will clamber over their neighbours. A good plant for heavy soils and part shade.

PLANT WITH

Verbena rigida

A low-growing perennial smothered in clusters of purple flowers June to September. Plant in a sheltered sunny spot in well-drained soil. A magnet for butterflies and bees.

H60cm x S40cm

Rosa
'Gertrude Jekyll'

Voluptuous bright-pink, repeat-flowering rose blooming from June to the frosts. Shrubs are medium-sized and upright in habit, preferring fertile, moist but well-drained soil.

H1.1m x S90cm.

Hakonechloa macra

Mid-green, shaggy-looking Japanese grass, with relaxed leaf blades that form cascading clumps at the border edge. Full sun or light shade and a moist well-drained soil.

H35cm x S40cm

H60cm x
S80cm

Sun or
Part Shade

Moist Well
Drained Soil

Pollinators

Frost Hardy

AGM Rated

Build Your Plant Matrix

Purple with...

+

+

+

+

+

+

+

+

Red

Geranium ‘Rozanne’ with red *Papaver commutatum* poppies (see page 22)

Pink

Salvia nemorosa ‘Caradonna’ with pink rose ‘Gertrude Jekyll’ (see page 62)

Purple

Salvia x *sylvestris* ‘Mainacht’ with *Nepeta racemosa* ‘Walker’s Low’

Blue

Crocus tommasinianus with reticulate iris ‘Harmony’ (see page 140)

Green

Allium ‘Purple Sensation’ with *Buxus sempervirens* (see page 275)

Yellow

Vinca minor with *Primula vulgaris* (see page 186)

Orange

Aquilegia vulgaris with orange tulips (see page 222)

White

Viola odorata with snowdrops (see page 74)

Blue

BLUE is the colour of cloudless skies and glittering oceans, birds' eggs and blue jeans. It's one of the three powerful primaries on the colour wheel, and within its cool embrace you'll find shades from dark, brooding indigo and navy to rich blues such as ultramarine and cobalt; mid-blues azure, cyan and cerulean; purple-blues such as lavender and lilac; and the glaucous green-blues: turquoise and teal.

Ultramarine was once one of the most expensive pigments on the artist's palette, made from the semi-precious stone, lapis lazuli. This is why in Renaissance paintings it's reserved for the cloak worn by the Virgin Mary, symbolizing holiness and humility.

True natural blues remain precious and hard to come by – especially when it comes to flower colour. In her book *Colour in the Flower Garden*, the venerable Victorian plantswoman Gertrude Jekyll drew up plans for a blue garden that excluded flowers with a purple-blue hue. She believed that pure blues, set off by white and pale-yellow partners, should never mingle directly with purple-blue ones. 'Surely the business of the blue garden is to be beautiful,' she wrote. 'Any experienced colourist knows that the blues will be more telling – and more purely blue – by the juxtaposition of a complementary colour.'

It's okay to disagree. The dearth of true-blue flowers means it's sometimes necessary to augment your planting palette with a few blue-purple blooms that will tone and harmonize. Certainly, a zing of yellow or white can prevent your blue border from looking too monochromatic; green foliage plants – ferns, hostas and strappy iris foliage – will also help to keep the blues and blue-purples apart.

Some of the best true blues belong to Himalayan poppies (*Meconopsis baileyi*); passion flowers (*Passiflora caerulea*); sages 'Oxford Blue' and 'Cambridge Blue'; cornflowers (*Centaurea cyanus*); forget-me-nots (*Myosotis sylvatica*); love-in-a-mist (*Nigella damascena*) and delphiniums. Campanulas and English bluebells (*Hyacinthoides non-scripta*) are a more purple-blue but could easily be worked into a blue border without too much soul-searching.

Silver foliage can also be very useful in fleshing out a blue planting scheme. Try senecio 'Silver Dust'; cineraria; curry plant (*Helichrysum italicum*); lamb's ear (*Stachys byzantina*); *Artemisia* spp; cotton lavender (*Santolina*); and blue fescue (*Festuca glauca*). They'll all enjoy a sunny hotspot and rub shoulders agreeably with crisp blue flowers. Some of the silver-leaved sub-shrubs have unremarkable yellow flowers that can quickly be snipped off with the secateurs should they offend.

In spring there are plenty of little blue flowers to choose from, contrasting attractively with dwarf narcissi. Choose from March-flowering scillas, ipheion and grape hyacinths, *Muscari armeniacum*. The bobbly flowered grape hyacinths are particularly easy and come in lots of exciting cultivars such as 'Peppermint', with its blue-and-white frosted look, catching the eye among evergreens and other spring groundcover plants.

Biennial forget-me-nots (*Myosotis sylvatica*) are another early starter and will quickly colonize a shady spot. They're such successful self-sowers you may eventually have to pull them out by the armful. If you like the little blue flowers but don't enjoy weeding, try perennial *Brunnera macrophylla* 'Jack Frost' instead; it has attractive heart-shaped silver-marbled green leaves and flowers April to May.

In late spring to early summer, it's the turn of blue camassias, which grow in the wild in damp meadows and woodland edges and will grow in damp heavy soils where many other bulbs would turn up their toes. Plant them en masse with other early perennials.

Summer brings the delphiniums – reportedly King Charles III's favourite flower in his garden at Highgrove, Gloucestershire. These regal-looking spires now come in a range of blues and whites that offer an exciting vertical accent among other sun-loving, cottage-style blooms. Annual larkspur, *Consolida* spp, is a wilder-looking alternative, often with single flowers that are favoured by bees.

In high summer (August to September time), your blue border will be electrified with the addition of *Agapanthus africanus*. These round-headed African lily flowers comprise trumpet-shaped tubes in a choice of blues, purples and whites, with straight stems and strappy foliage. Plant them in blocks or drifts, or in terracotta pots, or dot them through a border as repeated punctuation points that will draw the eye.

Late-summer-flowering salvias and eryngiums will take your blue border into autumn. Sometimes considered the bluest of the sea hollies, *Eryngium bourgatii* 'Picos Blue' and *E. zabellii* 'Big Blue' have large, intense, steely-blue flowers that thrive in a sunny spot.

In autumn, Michaelmas daisies such as 'Mönch' and 'Little Carlow' have lavender-blue daisies with a golden centre, blooming right to the frosts. They're resistant to mildew and attractive to wildlife, so there's no better way to see out the season. After flowering, cut down their flowered stems and apply a mulch, or leave it situ to create shelter for invertebrates, which in turn provide food for blackbirds and robins.

As a winter treat, plant a little pocket of blue dwarf irises. Reticulate cultivars such as 'Harmony', 'Alida' (a lovely grey-blue) and popular 'Katharine Hodgkin' (pale blue with a flash of yellow) will delight the eye when January starts to drag – growing best in a sunny, well-drained spot. Plant them among a few cute little evergreens, such as Juniper 'Blue Star' (H40cm) or *Festuca glauca* 'Intense Blue', for a stylish winter-proof display.

Muscari armeniacum

'PEPPERMINT'

These cute little grape hyacinths have flowers like bi-coloured bobble hats, often with a paler 'pompom' of florets at the top. Their grass-like foliage soon becomes a dense sward of green – useful ground cover at the edge of a path or under spreading shrubs. Dig up and divide congested clumps to help them spread around the garden. Flowers March to May in sun or part shade and moist, well-drained soil.

PLANT WITH

Viola odorata

Pretty purple-blue perennial with heart-shaped semi-evergreen foliage. The fragrant flowers February to March are edible.

H20cm x S30cm

Narcissus
'Tete à Tete'

Popular dwarf narcissus flowering March into April, with deep yellow blooms and narrow, strap-like green foliage.

H20cm x S5cm

Moss

Hummocks of green moss make an attractive backdrop for little flowers in spring. These evergreen plants can often be found around the garden in damp spots. Particularly effective growing on weathered stones or logs.

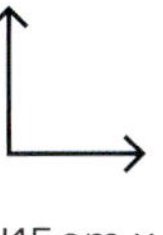

H15cm x S5cm

Sun or Part Shade

Moist Well Drained Soil

Pollinators

Frost Hardy

AGM Rated

Scilla siberica

Delicate-looking but tough, these nodding bell-shaped flowers have a pale centre and a blue stripe on each petal, flowering March to April. Known as the Siberian squill, each stem carries up to five flowers on darker stems, with narrow, strap-like foliage. They self-sow and naturalize to create attractive carpets in sun or dappled shade and a moist but well-drained soil.

PLANT WITH

Muscari armeniacum

Grape hyacinths with bi-coloured bobble flowers March to May. Good for naturalizing as ground cover in sun or part shade. Cultivar 'Cupido' shown is as cute as they come.

H15cm x S5cm

Anemone blanda 'White Splendour'

Perennial wood anemone with large white daisies on slender stems and feathery foliage, March to April. Needs sun or part shade and well-drained soil.

H15cm x S15cm

Euphorbia characias 'Silver Swan'

Small evergreen sub-shrub with white and green variegated foliage. Flowers have a red eye, March to May. Sap can be an irritant so wear gloves if cutting back.

H1m x S1m

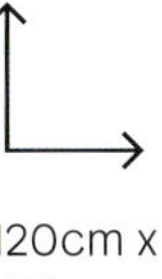

H20cm x
S5cm

Sun or
Part Shade

Moist Well
Drained Soil

Pollinators

Frost Hardy

AGM Rated

Hyacinthoides non-scripta

English bluebells spread to create a gentle haze of purple-blue flowers April to May. Their nodding flowers are all arrayed to one side of the stem – which is how you can tell them from their more vigorous Spanish cousin, *Hyacinthoides hispanica*. Only buy them from a licensed breeder as it's illegal to uproot them from the wild. They prefer a dappled shady spot under deciduous trees or shrubs in a humus-rich soil.

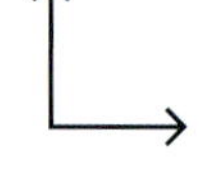

H40cm x S8cm

Sun or Part Shade

Moist Well Drained Soil

Frost Hardy

PLANT WITH

Myosotis sylvatica

Dainty little blue flowers float above clumps of soft green foliage April to June. A biennial self-sower that will quickly colonize a tricky dry shady spot.

H30cm x S15cm

Primula vulgaris

Native primrose with pale yellow flowers and a darker centre, March to May, above rosettes of crimped foliage. Good for naturalizing in light shade and a moist, well-drained soil.

H20cm x S35cm

Astelia chatamica

A clump-forming, half-hardy perennial with dramatic silver-green, strappy leaves and tall, pale green flowers June to August. Needs a sheltered spot in sun or part shade and may need protection in colder gardens.

H1.5m x S1m

Camassia leichtlinii

These starry blue florets emerge on stout flower spikes above strappy leaves from May. In the wild, camassias grow in damp meadows in North America – one of few perennial bulbs that likes moist soil (but not waterlogged). Choose a sunny or part shady position and partner with grasses, ferns and white umbelifers, such as *Anthriscus*.

PLANT WITH

Corydalis flexuosa
'China Blue'

Pretty perennial bearing clusters of dainty blue flower tubes April to May. Spreading mats of ferny foliage make this a useful groundcover plant for part shade under trees or shrubs.

H30cm x S20cm

Anthriscus sylvestris
'Ravenswing'

The billowing white umbels of native cow parsley emerge May to July, above clouds of feathery foliage. Best grown en masse for an ethereal effect in woodland-style borders.

H90cm x S60cm

Helianthemum
'The Bride'

Stunning, white-flowered sun-lover with grey-green to silver foliage and delicate blooms from May onwards. Drought tolerant.

H50cm x S50cm

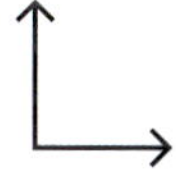

H80cm x S30cm

Sun or Part Shade

Moist Well Drained Soil

Frost Hardy

Ceanothus
'CONCHA'

Californian lilac is a large evergreen shrub with small, glossy, dark-green leaves and tightly packed clusters of blue flowers from May to June – adored by bees and other pollinators. Keep it looking trim by cutting back flowered shoots by one-third after flowering. Performs best in sun and a moist, well-drained soil.

PLANT WITH

Clematis alpina
'Helsingborg'

Vigorous climber with purple-blue bell-shaped flowers April to May and silvery, feathery seedheads. Look out also for cultivar 'Bredon Blue', which is bluer. Sun or light shade and moist, well-drained soil.

H2.5m x S1.5m

Viburnum plicatum f. *tomentosum*
'Kilimanjaro Sunset'

Bushy deciduous shrub with pleated leaves and horizontal tiered branches of white lacecap flowers April to June, followed by autumn berries. 'Kilimanjaro Sunset' is a charming cultivar.

H4m x S4m

Rosmarinus officinalis
'Miss Jessopp's Upright'

Easy to grow culinary evergreen with blue flowers May to June and soft, bushy, aromatic leaves. Plant in a sunny sheltered spot – trailing and prostrate forms available.

H2m x S2m

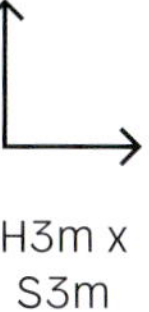

H3m x
S3m

Sun

Moist Well
Drained Soil

Pollinators

Frost Hardy

AGM Rated

Iris

'JANE PHILLIPS'

This sophisticated pale-blue bearded iris is highly rated by garden designers, appearing in RHS Chelsea Flower Show gardens from May to June. The complicated flowers – ruffled uprights, graceful falls with a hairy tongue – emerge on upright stems alongside fans of glaucous strappy foliage that stand well all year round. Perfect in a cottage garden or planted en masse in a sunny spot. Keep the rhizomes sitting at soil level – they enjoy being baked in the sun.

H1.2m x S60cm

Sun or Part Shade

Moist Well Drained Soil

Frost Hardy

AGM Rated

PLANT WITH

Salvia nemorosa
'Crystal Blue'

Perennial sage with tall spikes of sky-blue flowers June to October. Nectar-rich and fragrant for sun or part shade.

H45cm x S50cm

Anchusa azurea
'Loddon Royalist'

Intense blue flowers on tough stems with a bristly, hairy texture. Flowers June to August and needs full sun – loved by bees.

H1.5m x S60cm

Cynara cardunculus

Statuesque purple-blue flowered thistle with dramatic foliage, flowering June to September in full sun and moist but well-drained soil.

H2.4m x S90cm

Veronica

'SHIRLEY BLUE'

Perennial speedwell that forms low mats of grey-green foliage and spires of vivid blue flowers, May to July. Easy to grow, these long-flowering plants make attractive ground cover for the edge of a path or front of the border – a popular, wildlife-friendly choice for rockeries. It will perform best in full sun or light shade in moist well-drained soil.

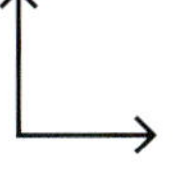

H25cm x S30cm

Sun or Part Shade

Moist Well Drained Soil

Pollinators

Frost Hardy

AGM Rated

PLANT WITH

Geranium
'Rozanne'

Reliable hardy geranium for sun or part shade, flowering June to October. Plants will clamber over their neighbours or spread along the ground – the flowers held on branching stems.

H60cm x S80cm

Geum
'Totally Tangerine'

Pretty, delicate-looking perennial with peachy-orange flowers on tall wiry stems. The sterile flowers last from June to September (they won't try to set seed so keep on going). Plant in sun or part shade and a moist, well-drained soil.

H90cm x S50cm

Convolvulus cneorum

Lovely silver-foliaged perennial, native to the Mediterranean. Pink buds open to white flowers, May to July. Plants need reliable sunshine and a well-drained soil.

H60cm x 75cm

Centaurea cyanus

These bright-blue hardy annual cornflowers really stand out in a 'cornfield-style' planting scheme, with yellow corncockle and red annual poppies. Loved by bees and other pollinators, the flamboyant little flowers appear on tall branching stems from May to July or July to September – depending on when you sow them. They prefer a sunny spot in poor soil and stems will be relatively lax in richer conditions eg clay.

PLANT WITH

Delphinium
'Loch Nevis'

Any blue delphinium will look fabulous in the company of annual cornflowers. This stunning hardy perennial cultivar makes a statuesque addition to a cottage-style border.

H1.5m x S30cm

Ammi majus

Dainty annual bishop's flower has lacy umbels and finely cut green foliage. Sow March to May for flowers in June to August. Best in a sun to part-shady spot in moist, well-drained soil.

H1m x S50cm

Festuca glauca
'Elijah Blue'

Silver-blue fescue forming neat clumps of pointy leaf blades and buff-coloured flowers June to July. Looks good with other silver foliage plants and those with blue flowers.

H30cm x S25cm

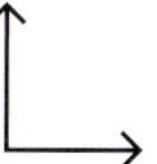

H70cm x
S30cm

Sun or
Part Shade

Moist Well
Drained Soil

Pollinators

Frost Hardy

Delphinium
'LOCH NEVIS'

Statuesque hardy perennial whose sumptuous double flowers have a white eye. Flowering June and July, these dramatic blue rockets are reported to be the King's favourite type of flower at his Highgrove Garden. They look best planted in odd-numbered groups, in a sunny sheltered site in well-drained soil that's rich and fertile. Best to stake them early and plant 60cm apart.

PLANT WITH

Eryngium x zabellii
'Big Blue'

Dashing sea holly with large, silver-blue, thistle-like flowers from June to August on tall wiry stems. Plants prefer a sunny, well-drained spot.

H1m x S60cm

Matthiola incana
'Beauty of Nice'

Fragrant cottage-garden biennial, forming leaves and roots in its first year, flowering the second. Grow them in sunshine from seed – available in a wide range of colours.

H50cm x S20cm

Senecio cineraria
'Silver Dust'

Tender sub-shrub with intricate lobed, velvety leaves, resistant to drought once established. Plant in sun in well-drained soil. Grow as an annual or in containers so you can move plants to a sheltered place in winter.

H30cm x S30cm

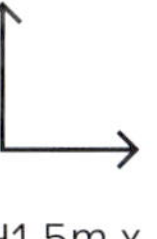

H1.5m x
S30cm

Sun or
Part Shade

Moist Well
Drained Soil

Frost Hardy

Agapanthus africanus

The African lily is an exciting addition to a blue border, flowering June to September. Its flowers resemble a controlled explosion of tubular blue trumpets held on stout, erect stems above strappy green foliage. They love a sunny spot in well-drained soil and quite like their bulbs to be a bit congested in a container. They may need frost protection in colder parts of the country where it's best to grow them in a container so you can bring them indoors over winter. Available in a wide range of blues and purples as well as white.

PLANT WITH

Echinops ritro
'Veitch's Blue'

Spiky globe thistle on tall stems, flowering July to August. It's a statuesque, pollinator-friendly perennial that thrives in sun or light shade and moist, well-drained soil.

H90cm x S45cm

Kniphofia
'Alcazar'

Bold orange poker, producing show-stopping 'look-at-me' blooms August to September above clumps of green foliage. Prefers a sunny spot in moist but well-drained soil.

H1.5m x S1m

Helichrysum italicum

Fragrant curry plant has silver-grey leaves and forms a dense, dwarf shrub with small yellow flowers July to August. Best in sun and moist, well-drained soil.

H60cm x S80cm

H60cm x
S45cm

Sun or
Part Shade

Moist Well
Drained Soil

Pollinators

AGM Rated

Meconopsis baileyi

Few flowers have the beauty and grace of these pure-blue, bowl-flowered Himalayan poppies. Flowering June to July on branching bristly stems, they have a reputation for being hard to grow and particular about their site and soil conditions – reliable moisture, acidic pH and light shade. They're fully hardy and sooo beautiful it's worth a try.

PLANT WITH

Centaurea cyanus

Bright-blue annual flowers on tall, wiry stems make this a must-have for any meadow planting scheme. Good for butterflies, May to September.

H70cm x S30cm

Meconopsis cambrica

Bright-yellow biennial (or short-lived perennial) Welsh poppy, flowering May to September. Ideal for a woodland border, plants will readily self-sow, preferring part shade.

H30cm x S45cm

Polystichum setiferum

Evergreen soft shield fern is perfect in a shady spot, unfurling new croziers from knobbly knuckles in spring. Cut back the old foliage in January.

H1.2m x S90cm

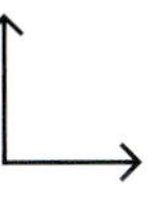

H1.2m x
S45cm

Part Shade

Moist Well
Drained Soil

Frost Hardy

AGM Rated

Campanula lactiflora

'PRICHARD'S VARIETY'

This cottage-garden perennial bears clusters of delicate bell-shaped trumpets on upright branching stems from June to September – a magnet for bees and other pollinators. Plant in sun or light shade in moist but well-drained soil. Cut back after flowering and divide larger clumps.

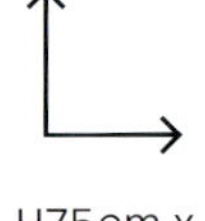

H75cm x S60cm

Sun or Part Shade

Moist Well Drained Soil

Pollinators

Frost Hardy

AGM Rated

PLANT WITH

Scabiosa caucasia

Soft purple-blue flowers are held on tall stems from July to September, above clumps of grey-green leaves. Intricate flowers can be deadheaded to encourage more. Cultivar 'Blue Diamonds' is shown here.

H60cm x S60cm

Leucanthemum vulgare

A charming native wildflower, the ox-eye daisy has large white flowers and a central golden disc, May to September. Grows best in nutrient-poor soil in sunshine.

H40cm x S30cm

Santolina chamaecyparissus

Silver-leaved cotton lavender forms pretty clumps of bobbly branching foliage. Cut it back hard in spring to keep plants neat, rather than woody and splaying in the centre. Ideal for a sunny, well-drained spot.

H50cm x S90cm

Nigella damascena

Love-in-a-mist is an evocative, ethereal-looking plant ideal for a cottage garden or blue planting scheme. Grow this hardy annual from seed – they come in a wide range of summer sky and midnight blues, and also white and pink. The intricate flower heads are borne on masses of green feathery foliage and produce intriguing seedheads that are every bit as lovely as the flowers – so don't cut them back too soon, and enjoy the self-sown offspring next summer. For sun and a moist well-drained soil.

H75cm x S50cm

Sun

Moist Well Drained Soil

Pollinators

Frost Hardy

PLANT WITH

Echinops ritro
'Veitch's Blue'

Purple-blue, spiky globe thistles on tall, silvery stems, loved by bees. Grow in well-drained soil in sun or part shade.

H90cm x S45cm

Eschscholzia californica

Zingy, citrus-orange annual Californian poppies grow on slender stems, with feathery foliage. They'll self-sow in paving cracks and gravel, flowering June to August.

H30cm x S15cm

Eryngium giganteum
'Silver Ghost'

Attractive silver-green biennial sea holly with a collar of spiky bracts around prickly flower cones, from June to August. Self-seeds freely in sun and well-drained soil.

H60cm x S40cm

Salvia patens

'CAMBRIDGE BLUE'

This gentian sage produces large, pale-blue hooded flowers and fresh green foliage from July to September. Ideal for a sheltered sunny spot in well-drained soil but not reliably hardy – you'll need to lift plants to store in frost-free conditions until replanting in spring or leave them in situ and cover with horticultural fleece. Apply mulch around the base in spring. 'Oxford Blue' is darker in hue.

PLANT WITH

Impomoea tricolor
'Heavenly Blue'

This morning glory cultivar has delightful blue and purple trumpet flowers from June to August. Plants are short-lived perennials grown as hardy annuals; sow them from seed in March or April, then plant out after all risk of frost in a moist but well-drained soil in sun.

H3m x 45cm

Echinacea purpurea
'White Swan'

A sophisticated white form of perennial echinacea, with gold-green nose cones and an upright habit. Flowers arrive July to October and are loved by bees and butterflies.

H60cm x S45cm

Taxus baccata

The glossy green foliage of this handsome evergreen can be trimmed into topiary shapes or planted as a low hedge to frame a parterre-style garden.

H10m x S6m

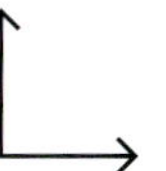

H75cm x
S50cm

Sun or
Part Shade

Moist Well
Drained Soil

Pollinators

AGM Rated

Eryngium x zabelii

'BIG BLUE'

H70cm x S60cm

Sun

Moist Well Drained Soil

Pollinators

Frost Hardy

AGM Rated

A spectacular sea holly with intense blue thistle-like flowers – a spiky thimble surrounded by a prickly ruff of bracts – perfect for a coastal-themed or dry gravel garden. Preferring a sunny position and well-drained soil, the flowers from August to October will attract bees and hoverflies looking for nectar. A bold statement plant with good seedheads that stand well in autumn and winter.

PLANT WITH

Lavandula angustifolia
'Hidcote'

Named after a famous garden in Gloucestershire, this fragrant sub-shrub draws in the bees. Purple-blue flower spikes July to September – prefers sun or part shade and moist well-drained soil.

H75cm x S60cm

Rosa
'Gertrude Jekyll'

Blowsy and beautiful, grow this rose in full sun and moist, well-drained soil. Be sure to mulch and prune annually while dormant.

H1.1m x S90cm

Artemisia ludoviciana
'Silver Queen'

Semi-evergreen perennial with slender silvery leaves and small yellow-brown flowers August to September. The foliage forms a low mound – perfect ground cover in a sunny spot.

H75cm x S60cm

Aster x *frikartii*

'MÖNCH'

A vigorous and disease-resistant Michaelmas daisy with lavender-blue flowers August to October on stout, branching stems. Plant in sun and a moist but well-drained soil for mounds of long-lasting flowers, each with a golden centre.

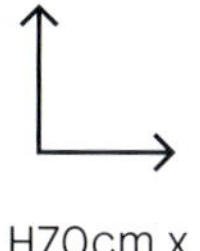

H70cm x S40cm

Sun

Moist Well Drained Soil

Pollinators

Frost Hardy

AGM Rated

PLANT WITH

Perovskia
'Blue Spire'

Russian sage with aromatic leaves and upright, silver stems bearing purple-blue flowers August to September. Good for a sunny position in dry chalky soil.

H1.2m x S1m

Solidago
'Babygold'

Goldenrod is a vigorous perennial ideal for prairie planting schemes and wildflower areas. Cultivars such as 'Babygold' and 'Goldkind' are smaller but just as attractive to butterflies, flowering July to August in sun and well-drained soils.

H60cm x S30cm

Stipa tenuissima

Wispy grass whose blades turn from green to bleached blonde, with fluffy flowers June to October. It's semi-evergreen and you can thin out congested clumps by combing through with your fingers.

H60cm x S30cm

Iris reticulata
'HARMONY'

Early flowering iris with striking, royal-blue flowers from January to February, each with a white throat and splash of yellow on its falls. Plant the bulbs 5cm deep in autumn in beds or borders in a sunny spot in moist but well-drained neutral to alkaline soil. They look good in containers too.

PLANT WITH

Iris reticulata
'Katharine Hodgkin'

Fancy-looking dwarf iris with blue veins and splashes of yellow on cream petals. Will grow well in sun or part shade, in pots or borders, attracting early bees in January and February.

H12cm x S8cm

Galanthus nivalis

Small, nodding winter flowers January to March, with distinctive green markings. Delicate honey fragrance. Grows best in sun or light shade around the ankles of deciduous trees and shrubs.

H10cm x S10cm

Helleborus x hybridus

The leathery evergreen palmate foliage of hellebores is an unsung hero of the shady border. It will act as a shapely foil for bright spring bulbs. Grow in moist, humus-rich soil.

H30cm x 45cm

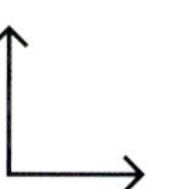

H15cm x
S8cm

Sun

Moist Well
Drained Soil

Frost Hardy

Iris reticulata

'KATHARINE HODGKIN'

This creamy-white, blue-veined reticulate iris is a delightful dwarf form, bearing large, distinctive flowers in March, each splashed with yellow and flecked with blue. Performs best in moist but well-drained soil in sun or part shade, flowering January to February.

PLANT WITH

Muscari armeniacum
'Peppermint'

Bobble-flowered miniature plant with grass-like foliage. Pale Delft-blue and white 'Peppermint' is a pretty cultivar to go for, flowering in March in sun or part shade.

H15cm x S5cm

Crocus chrysanthus
var. *fuscotinctus*

This unusual crocus has flashes of purple on the outside of its yellow goblet flowers. Naturalize in the lawn or borders for flowers February to March, with narrow, grass-like leaves.

H10cm x S10cm

Hedera helix
'Glacier'

Evergreen English ivy is a versatile companion for any winter flower, providing vigorous ground cover and covering ugly fences in just a few years. It's the most wildlife-friendly climber you can buy.

H2m x S2m

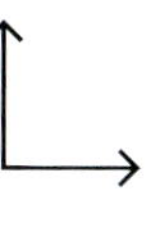

H12cm x S8cm

Sun or Part Shade

Moist Well Drained Soil

Pollinators

Frost Hardy

AGM Rated

Build Your
Plant Matrix

Blue with...

+

+

+

+

+

+

+

+

Red	*Festuca glauca* 'Elijah Blue' with red poppies (see page 22)
Pink	Bluebells with *Magnolia x loebneri* (see page 54)
Purple	*Delphinium* 'Summer Skies' with Geranium 'Rozanne' (see page 110) or campanulas (see page 134)
Blue	*Muscari armeniacum* with reticulate iris 'Harmony'
Green	*Omphalodes cappadocica* 'Starry Eyes' with *Helleborus foetidus* (see page 175)
Yellow	*Myosotis sylvatica* (forget-me-nots) with *Narcissus* 'Rip van Winkle (see page 184)
Orange	*Agapanthus africanus* with *Crocosmia* 'Emily McKenzie' (see page 32)
White	*Nigella damascena* in blue and white forms (see page 135)

Green

GREEN is the colour of plump spring buds, croquet lawns and fern-filled woodlands. It is the colour of leafy summer salads, shelled peas, lime cordial, peppermint and – love them or loathe them – Brussels sprouts.

The green colour of plants is created by the pigment chlorophyll, which helps them absorb energy from sunlight as they undergo photosynthesis, like mini solar panels.

Within the palette you'll find chartreuse, named after a French herbal liqueur, lime green and glaucous blue-greens. There's also moss, sage, olive, jade, emerald and dark forest green. Most of them harmonize surprisingly well.

Culturally, 'green' represents a broad cross-section of ideas: jealousy, naiveté, fertility and the environment. The notion that 'the grass is always greener on the other side of the fence' speaks not of our envy for the neighbours' perfect lawn, but rather, for the things that seem forever beyond our reach. Fortunately, a green-themed garden is something we can all achieve, and in our rain-tempered climate, quite easily.

Green has long been associated with our cultivation of the landscape. For the Romans, green was the colour of Venus, goddess of gardens, vegetables and vineyards (as well as sex and fertility). One only has to visit a patch of urban allotments to see the productive power of green in action, bringing people together in the pursuit of a beautiful, bountiful harvest.

Of course, rain helps. Wet, temperate regions like ours are always greener and more lush; this is something to be celebrated in our gardens by choosing diverse foliage plants that will create contrasts of shape, texture, hue and tone. Take inspiration from minimalist Japanese gardens, where they'll plant layers of green ferns, maples, moss and grasses.

This simple, limited palette is profoundly soothing – even therapeutic. Studies have shown that patients recover from surgery faster, and require fewer painkillers, if they can enjoy a tree-filled view from their hospital bed (Roger Ulrich et al 1984). It was also in the 1980s that the Japanese Ministry of Agriculture first coined the term *shinrin yoku*, or 'forest bathing', to describe the activity of exploring verdant surroundings for a calming effect on the frazzled human psyche.

In spring, the season begins with green buds and snowdrops nosing up through bare soil. Shade-loving hellebores are followed by viridiflora tulips and Soloman's seal (*Polygonatum* x *hybridum*), their contrasting flower forms lending to the unfolding drama as fresh, zesty-green herbaceous perennials emerge from a winter below ground.

As the season progresses, you can count on the wide-ranging *Euphorbia* genus to pick up this zingy colour scheme. One of the most diverse plant groups, many of the succulent types can cope with drought and will self-sow when conditions are right. Their sulphur-yellow flowers look very much at home in Mediterranean-style gravel gardens, and even tropical or cottage garden schemes. One the biggest and boldest is *E. characias* subsp. *wulfenii* – an architectural, billowing plant with enormous flower heads on stout woody stems, preferring a light soil In full sun. There's also marsh spurge *E. palustris*; colourful fireglow spurge *E. griffithii*; cushion spurge *E. epithymoides*; evergreen wood spurge *E. amygdaloides robbiae*; and sprawling *E. myrsinites*. There's one for every garden.

For the widest choice of green summer blooms, look to the half-hardy annuals. There are green nicotianas, zinnias, moluccella (bells of Ireland), dianthus, mathiasella, echinaceas and rudbeckias, whose seed can be bought for a few pounds a packet. They'll keep going into autumn if you deadhead them. Augment your autumn greenery with a lime-green pompon dahlia and pale-yellow sunflowers such as *Helianthus* 'Lemon Queen'. Foliage will play a key role – ferns, *Hakonechloa macra* and heucheras such as 'Lime Marmalade' will really help prolong the display.

Early winter is a time when evergreens come to the fore. There are dozens of lustrous, glossy-looking trees and shrubs that can shrug off frost and snow, their foliage providing a luxurious backdrop for herbaceous perennials all year round.

For a dense, green hedge, try sleek, waxy-leaved cherry laurel (*Prunus laurocerasus*), privet (*Ligustrum ovalifolium*) or shrubby honeysuckle (*Lonicera nitida*). Conifers such as dwarf mountain pine (*Pinus mugo*) and white cedar (*Thuja occidentalis*) can provide a strong, stoic anchor among the froth of seasonal flowers that come and go.

Topiary plants such as box (*Buxus sempervirens*) and its alternatives – thornless Japanese holly (*Ilex crenata*), *Euonymus fortunei* and yew (*Taxus baccata*) – can provide a parterre framework of low hedging, or whimsical punctuation points in cones, mounds and lollipops.

Don't forget there's a wide range of variegated shrubs and climbers to choose from – hollies, ivies and hostas in summer. They come in cool white-and-green (eg *Euonymus* 'Emerald Gaiety') and warmer gold-and-green (eg *Euonymus* 'Emerald 'n' Gold'). These distinctive foliage plants will quickly brighten up a part-shady spot.

Tulipa
'SPRING GREEN'

Chic, sophisticated, creamy-white tulip with elegant green stripes on its petals, flowering in May. Plant in drifts and blocks, or dot them among evergreen ferns and grasses for a smart but naturalistic look. Perfect for a white or green-themed border with the partners below. Best in a sunny spot but will cope with light shade, in moist but well-drained soil.

PLANT WITH

Euphorbia epithymoides

Distinctive, chartreuse-green, mound-forming perennial with open, starry flower clusters April to May. Prefers full sun and a moist well-drained soil.

H40cm x S60cm

Myosotis sylvcatica

Native forget-me-not carries dainty little blue flowers from April to June above clumps of lance-shaped green leaves. Vigorous and tolerant of shade and heavy soil.

H30cm x S15cm

Euonymus fortunei
'Emerald Gaiety'

Green-and-white variegated semi-evergreen shrub, which often keeps its winter foliage. Insignificant green flowers in May and June, preferring sun or light shade and a moist well-drained soil.

H1m x S1.5m

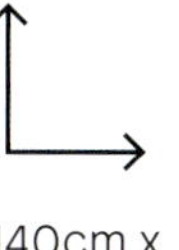

H40cm x
S10cm

Sun or
Part Shade

Moist Well
Drained Soil

Pollinators

Frost Hardy

AGM Rated

Euphorbia epithymoides

With its open, starry, chartreuse-green flowers above lance-shaped green leaves, this handsome spurge is a mound-forming perennial blooming April to May. It's great for creating rhythm in a herbaceous border and is a more interesting alternative to shrubs such as box or yew. Grow it mid-border or path side in full sun and a moist well-drained soil.

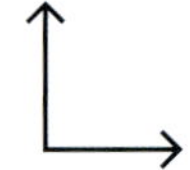

H40cm x S60cm

Sun

Moist Well Drained Soil

Pollinators

Frost Hardy

PLANT WITH

Doronicum x excelsum
'Harpur Crewe'

The bright-yellow daisies of leopard's bane have slender petals and heart-shaped, mid-green foliage. Their pert yellow centres are loved by bees – an early-flowering alternative (May to June) to rudbeckias and perennial sunflowers, which will cope with part shade.

H90cm x S60cm

Tulipa
'Ballerina'

Lily-flowered tulip producing bright, citrus-cocktail orange flowers in May. Lovely partner for acid-yellow flowers. Plant the bulbs 15–20cm deep in autumn, in a sunny spot in light, well-drained soil.

H55cm x S10cm

Phormium
'Yellow Wave'

Strappy-leaved, architectural perennial with yellow-green striped foliage. Sculptural evergreen presence in winter, though may need frost protection in cold winters.

Euphorbia characias var. *wulfenii*

This statuesque evergreen sub-shrub is a knock-out in springtime when its early, acid-green flowers start to emerge, March to July. Plants have a billowing, rounded habit, with stout, woody, sap-filled stems topped by enormous flowerheads, each with a burgundy eye. It's a big plant and will quickly dominate a small border where conditions are right, and will self-sow around the garden. Cut back flowered stems after flowering but wear gloves as the sap can irritate the skin. Prefers full sun.

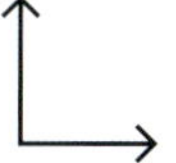

H1.2m x S1.2m

Sun

Moist Well Drained Soil

Pollinators

Frost Hardy

PLANT WITH

Nicotiana
'Lime Green'

Fragrant green-flowered tobacco plant flowering July to September. Plants are usually grown as an annual but can survive winter if mulched. Best in sun or light shade and a moist well-drained soil.

H75cm x S45cm

Digitalis purpurea

Native foxgloves will add height to a woodland-style border and their thimble-like flowers are loved by bumblebees, June to July. Plants are biennial, producing a rosette of leaves in their first year and prefer a shady position.

H2m x S60cm

Heuchera
'Lime Marmalade'

Lime-green foliage plant with ruffled, scalloped edges that make for pretty ground cover. Plants are semi-evergreen and produce sprays of creamy white flowers in June. Site in sun or part shade in moist well-drained soil.

H40cm x S30cm

Polygonatum x hybridum

Solomon's seal bears dangly, green-tipped white flowers on long arching stems – hanging like strings of dipped pearls from May to June. The foliage is lush and slightly ribbed, with a waxy underside. This striking perennial is ideal for a shady spot or woodland-type border, in a moist but well-drained soil.

PLANT WITH

Viburnum plicatum f. *tomentosum* 'Mariesii'

Japanese snowball bush is a deciduous shrub with large, lacy white flowers in May that resemble a lacecap hydrangea. The branches extend in horizontal layers creating an elegant effect that casts light-dappled shade. Plant in full sun or part shade in moist but well-drained soil.

H3m x S4m

Digitalis purpurea

Our pretty pink native foxgloves add height to a woodland-style border and are loved by furry-bottomed bumblebees when in flower, June to July. Plant these beautiful biennials in part shade in fertile, well-drained soil.

H2m x S60cm

Brunnera macrophylla

Large, heart-shaped leaves with silvery marbling set this reliable perennial apart. Great for ground cover, the leaves die back to black in winter, but return with new growth early in the season. Best in part shade – can scorch in full sun. Pretty blue flowers April to May.

H40cm x S60cm

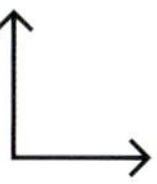

H1.5m x
S30cm

Part Shade or
Full Shade

Moist Well
Drained Soil

Frost Hardy

AGM Rated

Mathiasella bupleuroides

'GREEN DREAM'

Compact semi-evergreen herbaceous perennial with nodding, jade-green bracts each with a black flower centre, flowering from April to June. Flowers turn shell pink as the season progresses and last well into autumn. They're ideal for a cottage garden or woodland-style planting scheme, adding a distinctive architectural look. They're useful for cut flowers too. Plant in sun or part shade.

PLANT WITH

Angelica archangelica

Short-lived perennial with spherical green umbels June to July on robust, branching burgundy stems. Prefers sun or part shade and moist well-drained soil.

H2m x S1.2m

Cirsium rivulare 'Atropurpureum'

Hot pink, thistle-flowered perennial with serrated spiky foliage and a wild look, flowering July to August. Can cope with drought and dry shade.

H1.2m x S60cm

Physocarpus opulifolius 'Diabolo'

This devilishly handsome dark-leaved deciduous shrub has pink-tinted white flowers in June, but provides a dramatic foil for lime-green flowers throughout spring and summer. Plant in moist but well-drained acidic soil.

H2.5m x S2m

H1m x
S80cm

Sun or
Part Shade

Moist Well
Drained Soil

Pollinators

Frost Hardy

Aquilegia vulgaris var. *stellata*
'GREEN APPLES'

This cottage-garden 'granny's bonnet' has a demure look, with its intricate double flowers on upright stems from May to July. Flowers mature from green to crisp white, rising high above ferny foliage. Plants will self-sow in the right conditions – preferring sun or part shade in moist but well-drained soil.

PLANT WITH

Alchemilla mollis

Chartreuse-flowered perennial with gorgeous, scalloped leaves that hold raindrops like tiny balls of silver mercury. Sprays of tiny flowers June to September. Plants in sun or part shade.

H50cm x S50cm

Anchusa azurea
'Loddon Royalist'

Upright perennial bearing small, deep-blue, cup-shaped, bee-friendly flowers in June, maturing to purple. Plant in sun in a moist well-drained soil.

H1m x S60cm

Brunnera macrophylla
'Jack Frost'

Large heart-shaped leaves with silvery marbling set this reliable perennial apart. Great for ground cover, the leaves die back to black in winter, but return with new growth early in the season. Best in part shade – can scorch in full sun. Pretty blue flowers April to May.

H40cm x S60cm

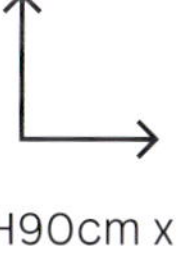

H90cm x
S40cm

Sun or
Part Shade

Moist Well
Drained Soil

Frost Hardy

Angelica archangelica

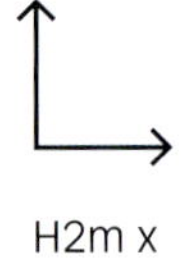

H2m x S1.2m

Green, sputnik-flowered short-lived perennial with spherical umbels from June to July on stout, branching stems that have a burgundy hue. Will self-sow in the right conditions – pot up the seedlings and move them where you want them to flower next year.

Sun or Part Shade

Moist Well Drained Soil

Pollinators

Frost Hardy

AGM Rated

PLANT WITH

Allium
'Mont Blanc'

Tall, white-flowered ornamental onion comprising slender green stems topped with spherical 'pompom' flowers. Repeat throughout a planting scheme to achieve a striking percussive effect. Prefers sun and a moist well-drained soil. Foliage can become tatty.

H1.1m x S20cm

Aquilegia vulgaris* var. *stellata
'Black Barlow'

A dark and dramatic contrast for white and green flowers, 'Black Barlow' is a fully double cultivar without spurs. A good self-sower that prefers sun or light shade and a moist but well-drained soil.

H90cm x S45cm

Foeniculum vulgare
'Giant Bronze'

Bronze fennel is a herbaceous perennial that produces clouds of light, feathery foliage and sulphur-yellow umbel flowers July to August, followed by sculptural seedheads. Plants are large and architectural, preferring full sun.

H1.8m x S45cm

Astrantia major

'SHAGGY'

Hattie's pincushion is an attractive clump-forming perennial flowering June to August on upright stems above a neat mound of mid-green, divided foliage. This cultivar has delicate white flower bracts tipped in green, with pink-flushed stamens. Plant in sun or light shade in a moist but well-drained soil. A self-sower when conditions are right.

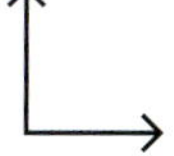

H80cm x S35cm

Sun or Part Shade

Moist Well Drained Soil

Pollinators

Frost Hardy

PLANT WITH

Digitalis lutea

Creamy-flowered perennial foxglove producing lots of flower stems from the base, May to July. Prefers a part-shady position in moist well-drained soil.

H60cm x S45cm

Geranium
'Brookside'

Blue-cupped geranium flowering June to September, with vigorous mounds of foliage that provide excellent summer ground cover. Prefers sun or light shade and moist well-drained soil.

H60cm x S45cm

Carex morrowii
'Everglow'

Bold green and cream-striped sedge forming neat clumps of colourful blades all year round. As temperatures drop the cream turns to a warmer orange hue. Grow in sun or light shade, in moist well-drained soil.

H60cm x S50cm

Alchemilla mollis

Chartreuse-flowered perennial forming a froth of starry little blooms June to September. The foliage is equally lovely: scalloped and soft to the touch, they'll hold onto dew and raindrops, turning them into tiny, shiny silver balls. Plant in sun or part shade in a moist but well-drained soil. A good self-sower when conditions are right.

PLANT WITH

Hydrangea arborescens
'Lime Rickey'

The luminous greenish white domed flowerheads of this deciduous hydrangea have lasting appeal from July to September. Easy to prune: just cut off the spent flowers after blooming. Best in light shade and a reliably moist but not waterlogged soil.

H1.2m x S1.2m

Brunnera macrophylla
'Jack Frost'

This easy perennial has heart-shaped foliage dusted with silver. Best in part shade – the leaves can scorch brown. Pretty blue forget-me-not-type flowers April to May.

H40cm x S60cm

Hakonechloa macra
'Aureola'

Handsome clump-forming Japanese grass that's loved for its texture as much as its colour. This striped golden form is a useful option to brighten a shady corner.

H35cm x S40cm

H50cm x S50cm

Sun or Part Shade

Moist Well Drained Soil

Frost Hardy

AGM Rated

Nicotiana

'LIME GREEN'

This long-flowering half-hardy annual produces fragrant, velvety, lime-green tube flowers that are pollinated by moths, July to September. By picking them for the vase you promote branching and encourage more flowers. Best to sow undercover in mid-spring, in a container. Plant out after all risk of frost has passed in late May.

PLANT WITH

Zinnia elegans
'Envy'

A must-have half-hardy annual flowering July to October, in sun or tolerating some shade. This cultivar has an AGM and is bee-friendly too. Sow outdoors from May onwards where they are to flower, thinning to 30cm apart, or sow undercover in trays of seed compost and plant out after the last frosts.

H60cm x S30cm

Verbena bonariensis

This stately, eye-catching perennial carries tightly packed purple flowers on branching wiry stems from June to October. It's a classic see-through plant, allowing you to use it towards the front of a planting scheme. Adored by bees and butterflies too. Best in sun and a moist but well-drained soil.

H2m x S45cm

Hordeum jubatum

The pink-tinted flowers of squirrel tail grass can really set off green-flowering plants, flowering June to July. Although it's a short-lived perennial, it is good at self-sowing, so should produce offspring around the garden.

H50cm x S30cm

H75cm x
S45cm

Sun or
Part Shade

Moist Well
Drained Soil

Pollinators

AGM Rated

Clematis florida var. *flore-pleno*

'PLENA'

Pretty, double flowers in creamy green-white that form rosettes from June to September. Foliage is semi-evergreen and needs training against wires, trellis or obelisk for support. Hard prune to 30cm the previous year's growth in March and top-dress with compost and bonemeal to feed the new flower-bearing growth.

PLANT WITH

Rosa
'Iceberg'

Clusters of pink rose buds open to double, slightly fragrant white flowers. Disease-resistant and easy to grow, flowering June to September.

H1.2 x S1m

Clematis
'Etoile Violette'

Purple-flowered clematis blooming July to September, it's an attractive option for trellis or an obelisk, or to clamber through a large shrub or tree.

H5m x S1.5m

Fatsia japonica

Sleek, glossy evergreen with large, deeply lobed leaves creating a modern, tropical look. Best in sun or part shade in a moist but well-drained soil.

H2m x S2m

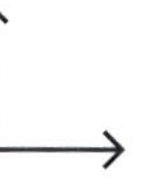

H4m x
S1m

Sun or
Part Shade

Moist Well
Drained Soil

Frost Hardy

Gladiolus

'GREEN STAR'

Ostentatious flower spikes of ruffled lime-green trumpet flowers lasting July to August. Growing from corms each year, plants are borderline hardy and will need protection in winter. Choose a position in sun or light shade in a fertile well-drained sandy soil.

PLANT WITH

Euphorbia characias var. *wulfenii*

Statuesque evergreen with acid-green flowers March to July. Plants have a billowing, rounded habit, with stout, woody, sap-filled stems. Prefers full sun.

H1.2m x S1.2m

Echinacea purpurea

Rose-pink daisies flower from June to September, with fuzzy, orange-brown nose cones loved by butterflies. Plant them in full sun for best results, in a moist well-drained soil.

H1.5m x S45cm

Taxus baccata

Glossy, evergreen, needle-leaved conifer available in columnar and golden forms, as well as the classic form that's ideal for hedging and topiary. Red arils form on female plants. A good backdrop for any herbaceous planting, also serving to anchor a cottage-garden scheme. Sun or shade.

Keep clipped or plants will grow to H12m x S8m+ in 50 years

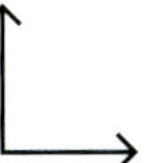

H1m x
S45cm

Sun or
Part Shade

Moist Well
Drained Soil

Zinnia elegans

'ENVY'

Zesty, green-flowered half-hardy annual blooming July to October, in sun or part shade. This AGM cultivar is bee friendly and easy to grow from seed. Sow outdoors from May where they are to flower, thinning to 30cm apart, or sow undercover in trays of seed compost and plant out after the last frosts.

PLANT WITH

Nicotiana langsdorffii

Citrus-green, trumpet-flowered half-hardy annual for sun or shade. Sow under cover in spring into trays of seed compost and cover lightly. Plant out after all risk of frost has passed for flowers July to October.

H1.5m x S45cm

Cosmos bipinnatus
'Sensation Dazzler'

These magenta pink flowers appear June to October on tall stems above a cloud of fine, feathery leaves. Sow these half-hardy annuals under cover March to May and plant out after all risk of frost.

H1.2m x S30cm

Hordeum jubatum

A short-lived perennial grass producing pink to green tufted flowers, June to July. A good self-sower when conditions are right, and a lovely way to introduce movement to a static planting scheme.

H60cm x S50cm

H60cm x
S30cm

Sun or
Part Shade

Moist Well
Drained Soil

Pollinators

AGM Rated

Humulus lupulus

'AUREUS'

Golden hop is an unusual climber for the garden, but its pretty chartreuse-green leaves and fragrant cone-like September flowers are ideal for a sunny or part-shady spot, clambering along a pergola, strong fence or house wall. Plants are deciduous, with best leaf colour in autumn. Cut back in spring to promote vigorous new growth.

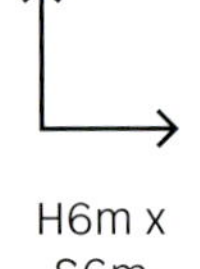

H6m x S6m

Sun or Part Shade

Moist Well Drained Soil

Frost Hardy

AGM Rated

PLANT WITH

Clematis florida var. *flore-pleno* 'Plena'

Creamy, green-white double, chrysanthemum-like flowers June to September. Foliage is semi-evergreen and needs training against wires, trellis or obelisk for support.

H4m x S1m

Clematis 'Etoile Violette'

Regal, purple-flowered clematis producing stunning blooms July to September, with pollen-dusted stamens attractive to bees. A good choice for a sunny or part-shady sheltered spot.

H5m x S1.5m

Fargesia murielae 'Luca'

Dwarf bamboo (H50cm) with vibrant evergreen foliage. Non-invasive and ideal for a container. For something a little taller and more dramatic, go for the straight species *Fargesia murielae* or exciting red-cane cultivar 'Red Panda'.

Both H4m x S1.5m

Skimmia x confusa

'KEW GREEN'

Evergreen male skimmia with thick, waxy, lanceolate green foliage in winter, producing white flowers from green-ish buds April to May. These useful shade-loving plants are generally neat and dome-shaped, requiring little pruning. As a male plant it can help to pollinate female, berry-producing skimmias nearby. Plant in moist, well-drained soil.

H1m x S1.5m

Sun or Part Shade

Moist Well Drained Soil

Frost Hardy

AGM Rated

PLANT WITH

Sarcococca confusa

Sweet box that produces small white flowers December to March packed with powerful fragrance. Shrubs are dense and evergreen, with glossy green foliage. Ideal for a shady spot in fertile, moist but well-drained soil.

H2m x S1m

Galanthus nivalis

Common native snowdrop with white nodding flowers and green markings January to March. Lift and divide congested clumps after flowering to help them spread.

H10cm x S10cm

Carex oshimensis 'Evergold'

Eye-catching gold-striped evergreen sedge that forms a neat clump of leaves that look good all year round. Ideal contrast partner for larger-leaved perennials such as ferns or hostas, offering a tactile element to planting schemes.

H50cm x S50cm

Chrysanthemum

'FEELING GREEN'

Citrus-green 'pompon' chrysanthemum with erect habit and aromatic foliage. Plant in full sun in a well-drained, fertile soil for flowers August to October. Flowers fade to a softer mint green as they develop into a ball. 'Froggy' is another good green cultivar.

PLANT WITH

Dianthus barbatus
'Green Trick'

Whimsical sweet William with tufty, lime-green flowers June to October. A short-lived cottage-garden perennial for moist but well-drained soil in sunshine.

H20cm x S30cm

Alchemilla mollis

Ground cover perennial producing a froth of acid-green starry flowers June to September. Soft, scalloped foliage turns dew and raindrops into molten balls of silver.

H50cm x S50cm

Phormium
'Yellow Wave'

An architectural evergreen bearing strappy, arching leaves in yellow and green stripes. Handsome anchor plant in a green-themed border. May need frost protection in a cold garden.

H2m x S2.5m

H75cm x
S45cm

Sun

Moist Well
Drained Soil

Frost Hardy

Galanthus elwesii

H30cm x S10cm

Sun or Part Shade

Moist Well Drained Soil

Pollinators

Frost Hardy

AGM Rated

So-called 'giant' snowdrop compared to *Galanthus nivalis*, these demure nodding flowers still have the same charm but are a bit beefier looking, flowering January to March. Ideal planted in naturalistic drifts and clumps in grass or borders, or in pots, flowers have a honey fragrance if you get right up close. The cross-looking green markings on cultivar 'Grumpy' create a bit of extra character.

PLANT WITH

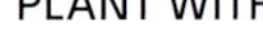

Helleborus niger

The Christmas rose makes an enchanting partner for any winter flowering shrub. Trim away the foliage in January so you can see the flowers better, January to February. A good choice for shade, with leathery evergreen foliage.

H30cm x S45cm

Eranthis hyemalis

Golden winter aconites, with their choir-boy collar of green bracts, make a cheerful partner for snowdrops, flowering January to February. Also good for naturalizing in part shade.

H8cm x S5cm

Hakonechloa macra

Mid-green, shaggy-looking Japanese grass, with relaxed leaf blades that form cascading clumps at the border edge. Full sun or light shade and a moist, well-drained soil.

H35cm x S40cm

Ribes sanguineum

'WHITE ICICLE'

This dainty-looking deciduous shrub is a real harbinger of spring, blooming from early March into April just when you need a bit of uplifting flower colour and fragrance. The layered branches produce a horizontal framework from which the pendant flowers dangle. Best in sun but will cope with part shade, in a moist, well-drained soil.

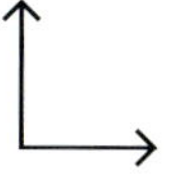

H1.5m x S1.5m

Sun or Part Shade

Moist Well Drained Soil

Pollinators

Frost Hardy

AGM Rated

PLANT WITH

Helleborus foetidus

Intricate bell-shaped, green flowers with a red rim mark out this distinctive hellebore, blooming from January to April. They're taller than many hellebore types and the handsome leaves give off an unpleasant odour if crushed. Sleek and good-looking in a shady 'woodland' border.

H80cm x S45cm

Narcissus 'February Gold'

Early daffodil that sets a cheerful tone for spring. Flowering from February in a mild winter until April, they perform best in a sunny spot in moist but well-drained soil.

H25cm x S10cm

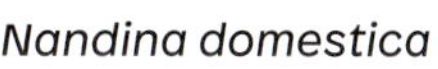

Nandina domestica 'Lemon and Lime'

Good-natured evergreen shrub that deserves to be grown more widely. Best in a sunny spot where its lime green foliage will retain its bright colour.

H60cm x S60cm

Build Your
Plant Matrix

Green with...

+

+

+

+

+

+

+

+

Red	*Helleborus argutifolius* with stems of *Cornus alba* 'Sibirica' (see page 16)
Pink	*Tulipa* 'Spring Green' with *Tulipa* 'China Pink' (see page 55)
Purple	*Nicotiana langsdorffii* with *Geranium* 'Brookside' (see page 59)
Blue	*Alchemilla mollis* with *Anchusa azurea* 'Loddon Royalist' (see page 124)
Green	*Euphobria characias* var. *wulfenii* with *Nicotiana* 'Lime Green'
Yellow	*Helleborus foetidus* with *Narcissus* *'February Gold'* (see page 184)
Orange	A Buxus or Taxus hedge surrounding *Fritillaria imperialis* 'Rubra' (see page 220)
White	*Brunnera macrophylla* 'Jack Frost' with *Narcissus* 'Thalia' (see page 250)

Yellow

YELLOW is the colour of summer sunshine and sandy beaches, bananas and custard, boiled eggs and buttered soldiers. It is the colour of seaside holidays: ice cream, fish and chips, buckets and spades. It is happy, positive, comforting and nostalgic.

We encounter this primary hue early in childhood – it's perfect for painting the sun's smiling orb and all manner of spring flowers, including dandelions and buttercups. It's the colour of daffodils on Mothering Sunday, fluffy Easter chicks, and the pollen-covered legs of furry-bottomed bumblebees.

It's also the colour of saffron, the most expensive spice in the world. Tricky to harvest, it's made from the stigmata of the *Crocus sativus* flowers. Fourteenth-century pirates preferred a bounty of saffron to gold because it was worth more, weight for weight, and was far easier to carry.

As a paint pigment, yellow has been derived from clay oxides – ochre, umber and sienna – since the earliest cave paintings, at Lascaux in France. In the more recent past, Claude Monet was among the first artists to exploit the bright, new, metal-based cadmium yellow pigment in his painting of *The Artist's House at Argenteuil* (1873). Vincent Van Gogh followed suit, using cadmium yellow in his painting of *The Yellow House* (1888) and his Sunflowers series (1887). Gustav Klimt uses 'cad yellow' too, in his glowing gold-leaf artwork, *The Kiss* (1908).

Given the enduring artistic caché of yellow, it's surprising the colour isn't more popular in the garden. It's often considered too bright and garish, although in February when we're desperate to move on from winter, any flash of yellow is gratefully received. Once the daffodils start appearing, we're happy to declare that spring has arrived, and start cleaning the house windows and dusting off the barbecue.

The other challenge with yellow is how to use it successfully with other colours. For instance, it can be difficult to combine warm yellow-variegated shrubs with cool white-variegated ones. Some might even say that yellow foliage looks a bit sickly or chlorotic (lacking in iron or manganese nutrients). In fact, healthy yellow leaves are coloured by natural pigments called xanthophylls, one of the carotenoid pigments, and they're meant to look like that.

Like it or not, yellow is the colour of springtime so it makes good sense to invest in a few of the main spring players – *Narcissus* being the favourites. Usually available at the garden centre as bulbs in autumn, look out for our wild native daffodil, *Narcissus pseudonarcissus*, or something more fancy such as double 'Rip van Winkle'. Early cultivar 'February Gold' will bloom slightly ahead of the others, and dwarf ones such as 'Minnow' and 'Tête-à-Tête' look cute and dinky under deciduous shrubs, on rockeries or in containers.

Yellow crocuses can be naturalized in the lawn in spring, at a time of year when it's a bit too early to start mowing. Pale yellow primulas are happy to naturalize too: *Primula vulgaris* is our native primrose, or 'first flower', blooming March to May; *Primula veris* is the taller one, the cowslip, flowering April to June.

For height, there are plenty of yellow-flowered spring shrubs. Among them, *Edgeworthia chrysantha* and *Forsythia x intermedia* both flower February to April; *Corylopsis paucifolia* and *Kerria japonica* both flower March to April. *Laburnum* is technically a tree, flowering May to June, but it can be trained on a pergola or metal arches to produce a stunning tunnel dripping with its pea-like flowers.

Summer brings yellow roses, such as *Rosa banksiae* 'Lutea' and 'Golden Wedding', blooming from June onwards. For mid-summer, sow packets of half-hardy annuals – bidens, tagetes, coreopsis and sunflowers – planting out your seedlings after all risk of frost has passed (usually late May).

Statuesque perennial verbascum can also be sown from seed with little fuss – just keep an eye out for wriggly mullein moth caterpillars: pull them off and pop them on the bird table. Soft yellow, cottage-garden cultivars such as 'Gainsborough' will reach 1.5 metres in height, flowering May to August. The flowers are worth the pest patrol.

Another good plant for soft, lemon-sorbet flowers is the perennial sunflower, *Helianthus* 'Lemon Queen'. These radiant, pale-yellow blooms on tall branching stems last from July into September. Partner them with long-lasting yellow dahlias, such as sputnik-flowered 'Honka', solidago and rudbeckia for gold flowers and seedheads well into autumn.

Don't forget foliage – golden evergreens are so useful in wintertime. Waxy-leaved *Choisya ternata* 'Sundance' will quickly brighten up a shady part of the garden and has breathtakingly fragrant blooms. Glossy evergreen *Mahonia* species can, between them, produce gold candelabrum flowers from November to March and beyond. 'Winter Sun' and 'Charity' are widely available and never disappoint. Plant them in a shrubby corner among fragrant witch hazels, such as *Hamamelis* 'Pallida' or 'Arnold Promise', for flowers December to February, surrounded by a carpet of winter aconites, *Eranthis hyemalis*. Golden conifers such as *Taxus baccata* 'Summergold' and *Pinus mugo* 'Winter Gold' will also pick up the colour baton, providing a dazzling year-round display in sun or shade.

Eranthis hyemalis

These cheerful, goblet-shaped flowers have a ruff of green bracts that act as a foil for the silky yellow flowers. In time they'll spread and naturalize to form bright carpets in a dappled-shady border or woodland planting scheme, flowering January to February. They're loved by those big, early, solitary bumblebees and tend to open when the sun comes out. Best in a fertile, moist but well-drained soil.

PLANT WITH

Iris reticulata
'Katharine Hodgkin'

Fancy-looking dwarf iris with blue veins and splashes of yellow on cream petals. Grows best in sun or part shade, in pots or borders, attracting early bees in January and February.

H12cm x S8cm

Galanthus nivalis

Elegant white, nodding winter flowers on slender green pedicels, with green markings and a sweet fragrance reminiscent of honey. Ideal for dappled shade under a deciduous tree or shrub.

H10cm x S10cm

Hamamelis
'Arnold Promise'

Handsome witch hazel with fragrant, whiskery yellow flowers on bare stems from January to February. Ideal for adding a bit of height and stature with spring flowers around its ankles.

H4m x S4m

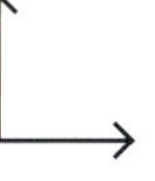

H8cm x S5cm

Sun, Shade or Part Shade

Moist Well Drained Soil

Pollinators

Frost Hardy

AGM Rated

Crocus chrysanthus var. *fuscotinctus*

Golden crocus native to the Balkans and Turkey, with fragrant, upright cup-shaped flowers that are smaller, and arrive earlier, than the Dutch crocus, *Crocus vernus*. These dainty bulbs will happily naturalize in a sunlit or dappled shady border, flowering February to March in a gritty, moist but well-drained soil.

PLANT WITH

Edgeworthia chrysantha

Deciduous shrub producing tight, rounded clusters of yellow and white flowers on bare stems February to April. Plant in a sunny, sheltered position and mulch well in spring.

H1.5m x S1.5m

Iris reticulata 'Harmony'

Early flowering iris with royal-blue flowers January to February, each with a white throat and yellow falls. Plant in beds or borders in a sunny spot in moist but well-drained soil, or containers.

H15cm x 8cm

Hedera helix 'Buttercup'

Evergreen ivy providing vigorous ground, wall and fence cover in just a few years. It's the most wildlife-friendly climber you can buy – providing lasting habitat, flowers and berries. This cultivar has yellow leaves and green leaves; for a variegated option, try 'Goldheart' or 'Goldchild'.

H2.5m x S2.5m

H8cm x
S8cm

Sun or Part
Shade

Moist Well
Drained Soil

Pollinators

Frost Hardy

Narcissus

'RIP VAN WINKLE'

This distinctive yellow narcissus is quite unique with its 'bed head' ruffled double flowers from March to April. They're a bit shorter and make a shapely contrast to more conventional daffs, drawing the eye. Plant your bulbs in autumn at about 10–15cm deep in a moist but well-drained soil in a sunny or part-shaded spot.

PLANT WITH

Narcissus
'February Gold'

Early daffodil that sets a cheerful tone for spring. Flowering from February, in a mild winter, until April, they perform best in a sunny spot in moist but well-drained soil.

H25cm x S10cm

Pulmonaria angustifolia

Purple to blue-flowered ground-cover plant, blooming March to May in sun or part shade. Stems are upright while the bell-shaped flowers gently dangle.

H30cm x S45cm

Choisya ternata
'Sundance'

Waxy-leaved evergreen shrub – Mexican orange blossom – that will lend the Midas touch to a part-shady spot all year round; its fragrant white flowers smell utterly delicious, April to May. Pruning can encourage a second flush, August to September. For moist, well-drained soil.

H2.5m x S2.5m

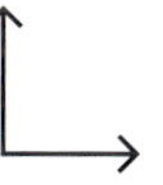

H14cm x
S8cm

Sun or
Part Shade

Moist Well
Drained Soil

Frost Hardy

Primula veris

Cowslips are a fabulous harbinger of spring that will last well into early summer, flowering April to June. It's hard to believe such a sensational flower, with its decadent clusters of yellow tubes, is so hardy and robust. They naturalize well in grass and are a good incentive to keep your lawn a little long to support visiting wildlife. Plant in sun or light shade, in moist but well-drained soil.

PLANT WITH

Primula vulgaris

Our native primrose – whose name means 'first flower' – blooms from March to May, with open cups of pale-yellow petals, each with a darker eye. They're perfect in the dappled shade of deciduous trees and shrubs in moist but well-drained soil.

H20cm x S35cm

Muscari armeniacum 'Peppermint'

Bobble-hatted grape hyacinth flowering March to May, best planted in sun or part shade and moist, well-drained soil.

H15cm x S5cm

Salix caprea 'Kilmarnock'

These pretty, pollen-dusted catkins belong to the Kilmarnock willow – a dwarf weeping tree whose bare branches carry male flowers designed for wind pollination. Plant in sun or shade in moist but well-drained soil.

H1.5m x S2.5m

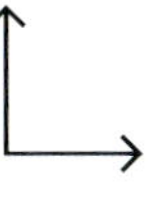

H25cm x S25cm

Sun or Part Shade

Moist Well Drained Soil

Pollinators

Frost Hardy

AGM Rated

Euphorbia amygdaloides var. *robbiae*

Zingy, robust and reliable, this chartreuse-coloured, evergreen wood spurge makes a bright addition to the spring garden, flowering April to June. Tolerant of dry shade and spreading by underground rhizomes, they're easy to pull up when they venture too far. The intricate flowers have cup-shaped involucres and nectar glands, loved by bees, hoverflies and other winged pollinators. The stout stems contain a milky sap that can be a skin irritant.

H70cm x S1m

Sun or Part Shade

Moist Well Drained Soil

Pollinators

Frost Hardy

AGM Rated

PLANT WITH

Tulipa
'West Point'

Large, lily-flowered yellow tulip with long, pointed, twisty petals, April and May. Lovely partner for euphorbia, alliums and *Erysimum*. Prefers sun and a well-drained soil.

H45cm x 10cm

Erysimum
'Bowles's Mauve'

Perennial, short-lived wallflower bearing masses of purple flowers February to July. Needs full sun for best results and a moist, well-drained soil.

H75cm x S60cm

Heuchera
'Lime Marmalade'

Lime-green, semi-evergreen foliage plant with scalloped edges and sprays of creamy white flowers in June. Site in sun or part shade in moist, well-drained soil.

H40cm x S30cm

Laburnum x watereri

'VOSSII'

Large, spreading 'golden chain' tree bearing long, dangling yellow flower racemes from May to June. The show-stopping blossom is loved by bees and followed by seeds that are poisonous if eaten. Bred in Holland in the 1880s, it has become one of the most popular cultivars, sometimes found trained into a laburnum tunnel – where plants are grown against a series of trellis arches. Little pruning is needed, but if you do need to curb its branches, prune while it's dormant in late summer to mid-winter as the sap can 'bleed'.

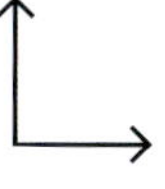

H5m x S4m

Sun or Part Shade

Moist Well Drained Soil

Pollinators

Frost Hardy

AGM Rated

PLANT WITH

Meconopsis cambrica

These silky, yellow Welsh poppies have a charming cottage-garden look, growing in path cracks and other garden crannies. Plants are biennial or short-lived perennials, easily grown from seed for flowers May to September. Plant in part shade and moist, well-drained soil.

H30cm x S45cm

Allium hollandicum
'Purple Sensation'

Upright purple spheres that strike a percussive note in any blowsy planting scheme. For flowers in May, plant the bulbs in autumn in sun and a moist but well-drained soil.

H1m x S1m

Hakonechloa macra
'Aureola'

Golden, shaggy-looking Japanese grass, with relaxed leaf blades that form cascading clumps at the border edge. Full sun or light shade and a moist, well-drained soil.

H35cm x S40cm

Achillea

'CLOTH OF GOLD'

These eye-catching umbellifers (yarrow) have densely packed yellow flowerheads that seem to float above the low clumps of feathery foliage, like golden platforms for pollinators. The tall stems emerge and flower June to September, but may need staking in rich soil and wet weather. Plant in full sun in moist but well-drained soil.

PLANT WITH

Rudbeckia fulgida var. *sullvianti* 'Goldsturm'

Flowering August to October, this handsome rudbeckia makes a lovely partner for achillea in a prairie-style border, among tall grasses. Plant in sun or light shade in moist, well-drained soil.

H60cm x S45cm

Salvia nemorosa 'Caradonna'

Popular perennial producing upright spires of rich purple florets, June to October. Cultivar *S. x sylvestris* 'Mainacht' flowers June to July, slightly bluer in colour, and taller at 75cm. Plant in sun and moist, well-drained soil.

H50cm x S30cm

Miscanthus sinensis 'Ferner Osten'

Clump-forming silvergrass bearing feathery flower plumes September to October. A lovely way to add movement to the autumn garden. 'Zebrinus' (H1.2m) is a stripy alternative.

H1.5m x S1m

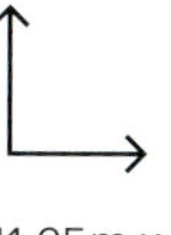

H1.25m x
S45cm

Sun

Moist Well
Drained Soil

Pollinators

Frost Hardy

AGM Rated

Hemerocallis lilioasphodelus

Flamboyant day lily with large trumpet flowers in butter yellow, flowering June. Plants are perennial and semi-evergreen in sheltered gardens, the flowers opening in the afternoon and lasting just one day. Perfect in pots and borders, in sun and a moist, well-drained soil.

PLANT WITH

Achillea
'Cloth of Gold

Startling yellow yarrow bearing dense umbels above low clumps of feathery foliage, June to September. Plant in full sun in moist but well-drained soil.

H1.25m x S45cm

Salvia x *sylvestris*
'Mainacht'

Clump forming perennial producing upright spires of blue-purple florets, June to July. Cultivar S. *nemorosa* 'Caradonna' flowers June to October, rich purple in colour, and shorter at 50cm. Plant in sun and moist, well-drained soil.

H75cm x S45cm

Hosta
'Whirlwind'

Clump-forming variegated hosta with ribbed foliage and pale lavender flowers July to August. The yellow variegation becomes lime green as leaves mature. Plant in shade and a moist, well-drained soil, sheltered from cold, drying winds, and protect from slugs with garlic spray.

H45cm x S1m

H1m x S30cm

Sun

Moist Well Drained Soil

Frost Hardy

Santolina rosmarinifolia

'LEMON FIZZ'

This unusual-looking, mound-forming perennial is ideal for a sunny sheltered spot, producing small, button-like yellow flowers on luminous yellow stems, July to August. With the common name of cotton lavender, it's perfect for a gravel garden or Mediterranean-style planting scheme. The foliage is aromatic when crushed. Plant in sun in a moist but well-drained soil.

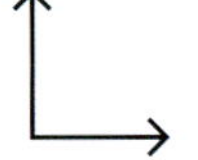

H50cm x S1m

Sun

Moist Well Drained Soil

Frost Hardy

AGM Rated

PLANT WITH

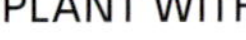

Alchemilla mollis

Pretty, ground-cover perennial bearing delicate sprays of acid-yellow-to-green flowers above scalloped foliage, June to September. Sun or part shade.

H50cm x S50cm

Lavandula angustifolia 'Hidcote'

Popular with pollinators, this compact English lavender carries fragrant purple flowers on silver-grey stems above aromatic foliage. Best in sun and a moist, well-drained soil.

H60cm x S75cm

Stachys byzantina 'Silver Carpet'

Velvet-leaved, silver-grey ground cover plant producing pink-purple flowers loved by bees June to August. Vigorous spreader when conditions are favourable – prefers sun and a moist, well-drained soil.

H20cm x S45cm

Helianthus

'LEMON QUEEN'

This lemon-yellow perennial sunflower, blooming July to September on branching plants, is quite unlike the more familiar annual sunflower, *Helianthus annuus*. Flowers are smaller but softer in colour with a dense golden nose cone, and make long-lasting cut flowers for indoors. Plant in sun and a moist, well-drained soil.

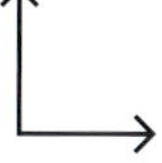

H2m x S45cm

Sun

Moist Well Drained Soil

Frost Hardy

AGM Rated

PLANT WITH

Verbascum
'Gainsborough'

A cottage-garden perennial bearing spires of cup-shaped, pastel-yellow flowers, May to August. Watch out for mullein moth caterpillars and pick them off by hand. For sun and a moist, well-drained soil.

H1.5m x S40cm

Aster x frikartii
'Mönch'

A vigorous and disease-resistant Michaelmas daisy flowering August to October. Plant in sun and a moist, well-drained soil for long-lasting flowers.

H70cm x S40cm

Cortaderia solloana
'Pumila'

Compact perennial pampas grass with silky flower plumes from August wafting above clumps of sharp-edged green foliage. A bold architectural focal point. Plant in full sun and moist, well-drained soil.

H1.5m x S1.2m

Lonicera periclymenum
'SCENTSATION'

Fragrant twining honeysuckle with large flowers in sunny, lemon-meringue colours, June to September. The blooms are attractive to pollinators, with clusters of shiny red fruits to follow. Prune after flowering if you need to contain its growth. Ideal way to hide a fence, providing nesting habitat for birds in the process. Plant in sun or part shade and a moist, well-drained soil.

PLANT WITH

Alchemilla mollis

Lovely billowing edging plant, with stays of bright yellow to green flowers June to September, held on tall stems above attractive green, scalloped foliage. Good self-sower in sun or part shade.

H50cm x S50cm

Rosa
'Gertrude Jekyll'

Robust shrub rose with fat pink flower buds that open to fragrant double blooms from June to September. Lovely for a romantic look in beds and borders, in sun and moist but well-drained soil.

H1.1m x S90cm

Stipa gigantea

A tall, bold, wafty grass creating movement in a static border. Clumps are semi-evergreen, with oat-like flowers on arching stems – hence the common name golden oats. Plant in sun and moist, well-drained soil.

H2.5m x S1.2m

H7m x
S1m

Sun or
Part Shade

Moist Well
Drained Soil

Pollinators

Frost Hardy

Clematis tangutica

The elegant nodding yellow flowers of this attractive clematis bloom from July to November before developing into distinctive puffball seedheads. 'Bill Mackenzie' is the best-known cultivar, with petals the texture of lemon peel. A good choice for a part-shady wall or fence in moist but well-drained soil. Prune hard in late winter.

PLANT WITH

Lonicera x brownii
'Golden Trumpet'

Eye-catching honeysuckle with vivid, golden-yellow blooms on twining stems June to August – preceding the blooms of *Clematis tangutica* – with red berries to follow. Plant in sun or part shade in moist, well-drained soil.

H4m x S2m

Clematis
'Perle d'Azur'

Soft, lilac-blue clematis flowering July to September. A lovely, vigorous companion for yellow clematis or roses. Plant in sun or part shade and cut back in early spring.

H3m x S1m

Cotinus coggygria
'Golden Spirit'

Gold-leaved deciduous smokebush, bearing smoky yellow flowers July to September, followed by fiery autumn tints. Best in sun, but also copes with part shade and moist, well-drained soil.

H5m x S5m

H4.5m x
S3m

Part Shade

Moist Well
Drained Soil

Pollinators

Frost Hardy

AGM Rated

Helenium

'EL DORADO'

Striking yellow-flowered 'sneezeweed', forming clumps of lance-shaped green foliage with erect flower stems emerging July to September. A pretty choice for a cottage garden or prairie-style border. Plants can tolerate most soil types but perform best in moist but well-drained conditions.

PLANT WITH

Inula helenium

These tall, feathery daisies with a brown eye make a shapely contrast for helenium and rudbeckia flowers, blooming June to August in sun and moist, well-drained soil.

H1.8m x S1m

Echinops ritro
'Veitch's Blue'

Spherical thistle-like flowers in steely blue-to-silver, July to August. Tall and architectural, it will draw in the bees and butterflies. Performs best in sun or light shade and moist, well-drained soil.

H90cm x S45cm

Hakonechloa macra
'Aureola'

Golden green, shaggy-looking Japanese grass, with relaxed leaf blades that form cascading clumps at the border edge. Full sun or light shade and a moist, well-drained soil.

H35cm x S40cm

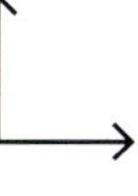
H75cm x S50cm

Sun or Part Shade

Moist Well Drained Soil

Pollinators

Frost Hardy

AGM Rated

Verbascum
'COTSWOLD QUEEN'

These dainty spires of cup-shaped, golden-yellow flowers have a cottage-garden look, flowering July to September. They're the perfect partner for other cottage-garden favourites, such as roses and clematis, but also work among tall grasses. The leaves of this perennial are semi-evergreen, so may offer ground cover in sheltered winter gardens. Watch out for mullein moth caterpillars – pick them off and pop them on the bird table. Best in full sun and a moist, well-drained soil.

PLANT WITH

Rudbeckia hirta
'Prairie Sun'

This annual black-eyed Susan has green eyes and daisy-like golden flowers July to October. Best sown from seed and grown as an annual, plants struggle to survive a British winter outdoors, but lovely to sow each year. Plant outdoors after all risk of frost.

H90cm x S45cm

Anemone hupehensis var. *japonica*
'Pamina'

Lipstick-pink Japanese anemone with double flowers July and August and deeply cut fresh green leaves. Vigorous, healthy and tolerates part shade.

H90cm x S40cm

Euonymus fortunei
'Emerald 'n' Gold'

Yellow-variegated evergreen shrub producing a dense, medium-sized bush with small green flowers May to June. Excellent backdrop or foil for contrasting plants or as part of a golden planting scheme. Easy to clip into a neat ball, it's an excellent, low-maintenance plant for sun or part shade and a moist, well-drained soil.

H60cm x S90cm

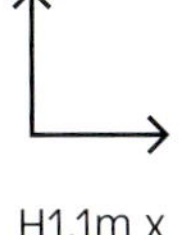

H1.1m x
S45cm

Sun

Moist Well
Drained Soil

Pollinators

Frost Hardy

AGM Rated

Kniphofia

'SUNNINGDALE YELLOW'

Dramatic, clump-forming, red hot poker that produces upright stems of sulphur-yellow tubular flowers July to October, above mounds of strappy foliage. With their rocket-like blooms, kniphofia are always an event when they flower – like dazzling fireworks. Plant your pokers in a sunny spot in moist but well-drained soil.

PLANT WITH

Helenium
'El Dorado'

Yellow-flowered, clump-forming perennial with erect flowers July to September. Plants are happy in most soils, preferring moist but well-drained conditions and sun or part shade.

H75cm x S50cm

Agastache
'Blue Fortune'

This handsome bee-magnet, giant hyssop, is a stately addition that's ideal towards the back of a sunny border, producing lofty spires of purple flowers June to October.

H1m x S40cm

Foeniculum vulgare

Yellow-flowered feathery herb whose foliage makes a fragrant and tactile contrast for other green or gold foliage and flowers. Blooms July and August, with shapely seedheads to follow. Plant in sun and moist, well-drained soil.

H1.8m x S45cm

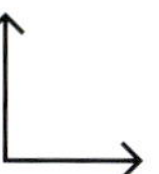

H1.15m x
S90cm

Sun

Moist Well
Drained Soil

Pollinators

Frost Hardy

AGM Rated

Rudbeckia fulgida var. *deamii*

Nicely proportioned, sunny-looking black-eyed Susan with dark nose cone, flowering August to October with attractive seedheads to follow. Looks lovely in a prairie-style border among grasses and other wild-looking golden flowers. Performs best in sun and a moist but well-drained soil.

PLANT WITH

Verbena bonariensis

Easy-going, pink or light purple-flowered perennial whose tightly packed flower clusters are held on branching, wiry stems. Its sparse silhouette means you can plant it at the front of a border and still see what's behind. Beautiful and useful.

H2m x S45cm

Verbascum 'Cotswold Queen'

Spires of golden, cup-shaped flowers with a purple boss of stamens, flowering July to September. Producing large, semi-evergreen leaves, you'll need to keep an eye out for mullein moth caterpillars. Best in full sun and a moist, well-drained soil.

H1.1m x S45cm

Calamagrostis x acutiflora 'Karl Foerster'

Upright, clump-forming feather-reed grass that forms an attractive screen for wheelie bins! Cut back to the ground in February – this grass is a deciduous perennial, so this will make space for new green leaf blades. Flowers June to August. Plant in sun or light shade in moist but well-drained soil.

H1.8m x S60cm

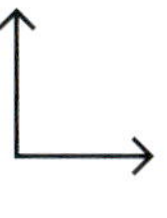

H60cm x
S45cm

Sun

Moist Well
Drained Soil

Pollinators

Frost Hardy

AGM Rated

Sternbergia lutea

Known as the autumn daffodil, this autumn-flowering perennial bulb blooms from September to November, with golden goblets that will naturalize and form carpets in time. Native to the eastern Mediterranean, it prefers a frost-free, sheltered, sunny spot in well-drained light, sandy soil.

PLANT WITH

Clematis tangutica

Elegant clematis with nodding yellow flowers July to November that develop into puffball seedheads. 'Bill Mackenzie' is the best-known cultivar. For part-shade and moist but well-drained soil. Prune hard in late winter.

H4.5m x S3m

Cyclamen hederifolium

Hardy autumn-flowering cyclamen with attractive ivy-like leaves and dainty pink or white blooms October to November. Looks good under deciduous trees and shrubs in a woodland-style border. Plant in humus-rich soil in part shade.

H10cm x S15cm

Carex elata 'Aurea'

Large, native, gold-striped evergreen sedge that forms large, rounded clumps ideal for ground cover among ferns and large-leaved perennials such as hostas. Other gold-bladed sedges include 'Everillo' and 'Eversheen'. Plant in sun or part shade in moist but well-drained soil.

H75cm x S75cm

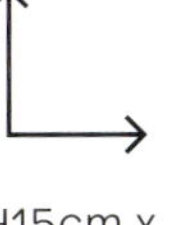

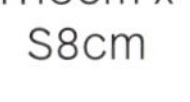

H15cm x
S8cm

Sun

Moist Well
Drained Soil

AGM Rated

Mahonia x media

'CHARITY'

An upright evergreen shrub with glossy, spiky foliage and candelabra-style yellow flowers November to March, followed by purple berries (not edible) that give this shrub its common name, the Oregon grape. Plants can be grown in sun or shade in a moist, well-drained soil. Prune after flowering and mulch with well-rotted organic matter.

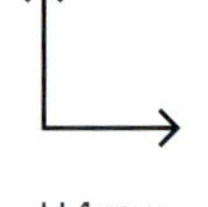

H4m x S4m

Sun or Part Shade

Moist Well Drained Soil

Pollinators

Frost Hardy

PLANT WITH

Hamamelis x intermedia 'Pallida'

Sweetly fragrant witch hazel producing ribbon-like, lemon-yellow flowers December to February, on bare branches. Statuesque partner for any tall winter shrub.

H3m x S3m

Crocus tommasinianus

Open, goblet-shaped purple flowers that are lovely naturalized in lawn – especially when planted en masse, flowering February to March. For dappled shade and a moist, well-drained soil.

H10cm x S30cm

Taxus baccata

Golden, columnar form of evergreen yew. Very smart in a formal garden setting or small 'avenue' and can be clipped into topiary forms. For sun or shade in moist, well-drained soil.

H5m x S1.5m in 20 years

Jasminum nudiflorum

Winter jasmine is a popular shrub for training against a wall – it can't twine like summer-flowering *Trachelospermum jasmine*, but it does have a unique appeal with its mass of yellow flowers on bare green stems, January to March. Plant in sun or part shade and a moist, well-drained soil.

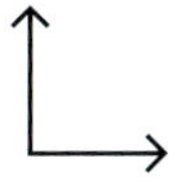

H2.5m x S2.5m

Sun or Part Shade

Moist Well Drained Soil

Frost Hardy

AGM Rated

PLANT WITH

Cornus mas

The Cornelian cherry offers reliable golden flowers on bare stems from January to February. Its leaves turn reddish-purple in autumn, with cherry-like fruits. Plant in sun or part shade in moist but well-drained soil.

H4m x S4m

Chimonanthus praecox

Wintersweet is a deciduous shrub bearing fragrant yellow flowers December to February, each splashed with a purple interior. Plant in a sunny spot in moist but well-drained soil.

H4m x S2.5m

Pinus mugo 'Winter Gold'

Characterful, dwarf pine tree with evergreen golden needles. Makes a lovely, bushy foliage contrast for winter flowering plants. Prefers sun and a well-drained soil.

H1.5m x S2m

Build Your
Plant Matrix

Yellow
with…

+
+
+
+
+
+
+
+

Red

Daffodils with stems of *Cornus alba* 'Sibirica' (see page 16)

Pink

Carex 'Everillo' with *Bergenia cordifolia* (see page 48)

Purple

Alchemilla mollis with *Geranium* 'Rozanne' (see page 110)

Blue

Helenium 'El Dorado' with *Aster* x *frikartii* 'Mönch' (see page 139)

Green

Euphorbia characias subsp. *wulfenii* with *Nicotiana* 'Lime Green' (see page 162)

Yellow

Helianthus 'Lemon Queen' with *Verbascum* 'Gainsborough'

Orange

Hemerocallis lilioasphodelus with *Crocosmia* 'Emily Mckenzie' (see page 32)

White

Forsythia x *intermedia* with *Narcissus* 'Thalia' (see page 250)

Orange

Orange is the colour of summer fruit and autumn leaves; chunky marmalade, carrots, bell peppers and pumpkins. It is the colour of Tigger, traffic cones and butternut squash; wily foxes, small copper butterflies and startled goldfish, darting around the shady shallows.

Orange shouts, in a bold, creative way. It is strident, but also mystical, exotic and sun-blushed, associated with warmer climates than our own. Orange flowers look resplendent beside purple and blue, and although timid gardeners may shy away from such a brazen hue, those who dabble rarely regret it. An orange kniphofia coming into flower is an Event.

Created by mixing red and yellow on the colour wheel, orange is a secondary hue that's surprisingly common in nature – especially in autumn when the green chlorophyll in leaves breaks down to reveal orange and yellow carotenoid and xanthophyll pigments. Within its range, you'll find pale sandy orange through to citrussy-satsuma, blushing blood orange, pinkish apricot, coral and darker tones of bronze, amber and rusty steel.

In art, the discovery of lead chromate (crocoite) in 1797 by French scientist Louis Vauquelin led to the invention of the synthetic chrome-orange pigment in 1809. This enabled artists such as Auguste Renoir and Toulouse Lautrec to produce far stronger effects than by mixing yellow and red pigments alone. Famous examples of orange in art include the Pre-Raphaelite painting *Flaming June* (1895) by Lord Frederic Leighton, showing a female model curled up on a chair in a beautifully draped orange dress. The Impressionists loved the colour too – finding its most vivid expression in relation to azure blue. 'There is no orange without blue,' wrote Van Gogh to his brother, referring to the way contrasting colours serve to set each other off, visually. It's great advice for planting schemes too.

Spring opens with the bee-friendly goblets of crocus 'Orange Monarch' and swiftly progresses with the orange trumpets of narcissus 'Golden Dawn' and 'Orangery', which emerge alongside orange wallflowers such as 'Apricot Twist'. But it's the tulips, such as lily-flowered 'Ballerina' and more rounded 'Prinses Irene', that really deliver the visual gusto. More exotic still are the startling flowers of *Fritillaria imperialis* 'Rubra', which bloom March to May on purple stems, their orange bells dangling below a crown of erect green bracts.

For early summer, you can't beat a sowing of neon-bright, hardy annual Californian poppies – *Eschscholzia californica*. Flowering on upright stems June to August above ferny, glaucous foliage, they'll quickly create a dazzling clump or drift that looks especially charming in gravel.

Best known of the orange summer flowers is probably the dainty perennial *Geum* 'Totally Tangerine'. Its perky, double, sunset-orange flowers on tall, fuzzy stems are sterile, so they last for months without trying to set seed, from June to September in sun or light shade, making this a firm favourite with garden designers.

Other long-lasting orange blooms include hardy annual pot marigold *Calendula officinalis*, as well as the half-hardy annuals, *Cosmos sulphureus* and Mexican sunflower *Tithonia* 'Torch'. A combination of these hot-flowered beauties, grown from seed to save cash, will keep the orange flickering through your planting scheme from June into autumn.

Midsummer brings the Asiatic lilies – their large, trumpet-shaped flowers adding a voluptuous, jungly note to any hot border. Orange cultivars include 'African Queen', 'Orange County' and 'Sunset Boulevard'. There are speckled tiger lilies too, *Lilium lancifolium*, and strap-leaved daylilies, *Hemerocallis fulva*. Plant them with canna lily 'Wyoming', *Crocosmia* 'Emily McKenzie', dahlia 'David Howard' and orange hot pokers such as *Kniphofia* 'Tawny King' for an indulgent array of exciting flower forms that will last well into October.

One exciting idea, pinched from Monet's garden in Giverny, is to sow orange-flowering nasturtium seed, *Tropaeolum majus*, along opposite sides of a gravel path and watch as they gradually reach across and meet in the middle, July to September.

Deciduous leaves, flamboyant dogwoods and the rusted seedheads of *Hylotelephium* will provide a range of warm autumn hues. Go for *Acer palmatum* cultivars such as 'Orange Lace' and 'Katsura' for reliable orange foliage as temperatures dip, underplanted with some rusty Buckler ferns – *Dryopteris erythorosa* – late-flowering dahlias and more kniphofia.

Don't forget that oranges are not the only fruit. Even in the depths of winter we can enjoy orange berries outside – from eye-catching *Pyracantha* cultivars such as 'Orange Glow' to *Ilex verticillata* 'Winter Gold' and *Sorbus* spp. Other winter treats include fragrant *Hamamelis* 'Jelena', whose spidery orange flowers resemble clusters of wrinkled orange peel on bare branches from January to February.

Narcissus
'GOLDEN DAWN'

Bright, orange-trumpeted tazetta-type daffodil with contrasting butter-yellow petals, flowering March to April. Plant the bulbs approx. 10–15cm deep in autumn in naturalistic drifts, in sun or part shade and moist, but well-drained soil.

PLANT WITH

Viola cornuta
'Penny Orange'

Delightful pale-orange viola with black 'whiskers' pointing to its nose. Flowers March to October – keep deadheading to encourage more. A useful bedding plant for sun or shade and a rich soil.

H15cm x 20cm

Scilla siberica

Dainty, nodding spring flower with violet-blue flowers March to April and grass-like foliage. For sun or part shade and moist, well-drained soil.

H20cm x S5cm

Dryopteris affinis
'Cristata'

Native woodland fern producing clumps of lush, semi-evergreen foliage. Prefers part shade and a moist, well-drained soil.

H90cm x S90cm

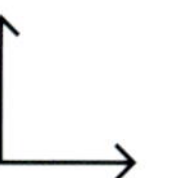

H40cm x S10cm

Sun or Part Shade

Moist Well Drained Soil

Erysimum cheiri
'APRICOT TWIST'

Sumptuous wallflower bearing bold, dark-orange flowers April to June, ideal for a cottage-garden border. This short-lived evergreen perennial can be grown as a biennial – that is, discarded after flowering. They make excellent partners for tulips in spring but tend to get a bit rangy in time. Plant in sun in moist but well-drained soil.

PLANT WITH

Tulipa
'Ballerina'

Lily-flowered orange tulip blooming April to May on tall, sturdy stems. Lovely among green and purple foliage plants. Performs best in a sunny spot in moist but well-drained soil.

H55cm x S10cm

Erysimum
'Bowles's Mauve'

Good-looking wallflower with bold purple blooms, February to July, and clumps of dark blue-grey lance-shaped leaves. Short-lived evergreen perennial that performs best in sun and a moist, well-drained soil.

H75cm x S60cm

Skimmia japonica
'Rubella'

Glossy-leaved evergreen shrub forming a neat, rounded shape. Red autumn buds open to white flowers April to May.

H1.5m x S1.5m

H50cm x
S50cm

Sun

Moist Well
Drained Soil

Pollinators

Frost Hardy

Fritillaria imperialis
'RUBRA'

This dramatic crown imperial is guaranteed to turn heads. These startling perennials bloom March to May on dark purple stems, their orange bells dangling below a crown of erect green bracts. Plant 20cm deep in autumn in a sunny spot with a moist well-drained soil.

PLANT WITH

Erysimum
'Rysi Copper'

Charming perennial wallflower with pale orange flowers March to October. A magnet for bees and butterflies – good with tulips too. Sun or part shade in moist, well-drained soil.

H50cm x S50cm

Tulipa
'Burgundy'

Dramatic, fluted, reddish-purple tulip flowering April to May on tall, upright stems. Other good dark ones are 'Havran' and black 'Queen of Night'. For sun or part shade and moist, well-drained soil.

H50cm x S10cm

Euonymus japonicus
'Green Spire'

Small-leaved, glossy, evergreen shrub that's a very smart and useful alternative to box, without the caterpillars or blight. Plant in sun or part shade in moist but well-drained soil.

H1m x S50cm

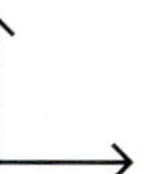

H70cm x
S20cm

Sun

Moist Well
Drained Soil

Frost Hardy

AGM Rated

Tulipa
'BALLERINA'

Elegant, elongated, lily-flowered tulip with outer petals flushed darker orange, April to May. Perfect drifted through purple and blue planting schemes, or yellow or green: a useful all-rounder that's just as good in a pot. The tall, sturdy stems make this a robust option that performs best in a sunny spot in moist but well-drained soil.

PLANT WITH

Fritillaria imperialis
'Rubra'

Dramatic crown imperial blooms March to May, their orange bells dangling below a crown of green bracts. Plant 20cm deep in autumn in sun and a moist well-drained soil.

H70cm x S20cm

Euphorbia characias subsp. *wulfenii*

Large, architectural gem for a big border, bearing whopping great acid-yellow flowers March to July. Prefers sun or part shade and moist, well-drained soil. Tolerant of drought once established.

H1.2m x S1.2m

Carex buchananii

Perennial sedge whose coppery blades team perfectly with orange flowers. Forms dense tussocks and small summer flowers. Plant in sun or part shade in moist but well-drained soil.

H60cm x S90cm

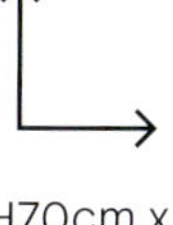

H70cm x
S20cm

Sun

Moist Well
Drained Soil

Frost Hardy

AGM Rated

Helenium
'SAHIN'S EARLY FLOWERER'

A stunning daisy-like perennial with orange petals arranged around bobbly, brown, pollen-dusted centres – hence the common name sneezeweed. Heleniums are utterly lovely in a meadow or cottage-garden-style planting scheme, flowering June to August. They prefer a sunny position in moist but well-drained soil.

PLANT WITH

Kniphofia
'Tawny King'

Orange-flowered poker adding height to a hot border from July to October. The torch-like blooms are an event to behold, exploding from the dense clump of strappy foliage. Best in a sunny spot or light shade, in a moist, well-drained soil.

H1.25m x S1m

Aster x frikartii
'Mönch'

Handsome, bushy Michaelmas daisy flowering August to October with large showy flowers loved by butterflies. Plant in sun or light shade and a moist, well-drained soil.

H70cm x S45cm

Dryopteris erythrosora

Semi-evergreen Buckler fern, whose young foliage has a bronze hue. Plants form neat clumps and will quickly create a woodland look in a shady spot in moist, fertile soil.

H1m x S1m

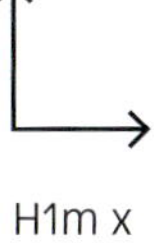

H1m x S50cm

Sun

Moist Well Drained Soil

Pollinators

Frost Hardy

AGM Rated

Eschscholzia californica

'ORANGE KING'

A classic orange poppy whose bold cocktail-coloured flowers, June to August, glow almost neon on an overcast day. These hardy annuals look gorgeous emerging through gravel, their ferny, green-blue foliage creating a frothy carpet. Grow them from seed, sowing March to May where they are to flower. Allow their seedheads to self-sow to ensure a continuing display. Prefers sun and a moist, well-drained soil.

PLANT WITH

Calendula officinalis

Versatile hardy annual, flowering June to October. Sow from seed March to May, then plant out in sun or part shade in moist, well-drained soil.

H60cm x S45cm

Asphodeline lutea

The yellow spires of perennial king's spear will thrive in a dry, sunny spot. Flowers June to July, rising from clumps of grey-blue grassy leaves. Best in sun and a moist, well-drained soil.

H90cm x S30cm

Stipa tenuissima

Wispy perennial grass forming relaxed clumps that move with the breeze. Flowers June to October. Comb through in spring to remove dead stems.

H60cm x S30cm

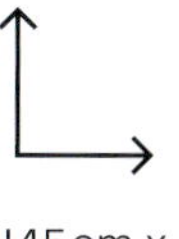

H45cm x
S30cm

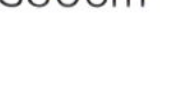

Sun

Moist Well
Drained Soil

Pollinators

Frost Hardy

Euphorbia griffithii
'FIREGLOW'

This good-looking, bushy perennial has all the robust reliability and versatility of the euphorbia family, with exciting fiery-orange-red flowers, June to September. Very much at home in a hot-border planting scheme, team it with other orange blooms or brooding purple flowers and foliage. Plant in sun or part shade and a moist, well-drained soil.

PLANT WITH

Achillea
'Terracotta'

The warm summer colours of this upright umbel make it the perfect partner for other orangey-yellow blooms, the hue mellowing as it matures, from June to September. Ideal in a cottage or prairie-style planting scheme in sun and a moist, well-drained soil.

H1.1m x S40cm

Allium
'Purple Sensation'

These upright, peppy purple spheres will provide percussive visual contrast for daisies, umbels and bushy euphorbia. Plant the bulbs in autumn in sun and a moist but well-drained soil.

H1m x S1m

Carex comans
'Bronze Form'

Copper-coloured sedge forming dense tussocks, with small summer flowers. Plant in sun or part shade in moist but well-drained soil.

H20cm x S30cm

H80cm x
S50cm

Sun or
Part Shade

Moist Well
Drained Soil

Frost Hardy

Geum

'TOTALLY TANGERINE'

Charming, popular, cottage-garden perennial bearing delicate-looking peachy-orange flowers on tall, wiry stems, above clumps of scalloped foliage. The flowers last for ages – June to September – thanks to them being sterile (they won't try to set seed). Plant in sun or part shade and a moist, well-drained soil.

PLANT WITH

Achillea
'Walther Funcke'

Upright umbel whose orange flowers mellow to yellow as it matures, from June to September. Ideal in a cottage or prairie-style planting scheme in sun and a moist, well-drained soil.

H60cm x S60cm

Cirsium rivulare
'Atropurpureum'

Ornamental magenta-flowered thistle with tufty blooms, tall branching stems and spiky leaves. Flowers July to August and prefers a sunny position in moist, well-drained soil.

H1.2m x S60cm

Polystichum setiferum

Evergreen soft shield fern has dark-green fronds that unfurl into a spreading array of tactile foliage. A shade lover for moist but well-drained soil. Cut back in early January.

H1.2m x S90cm

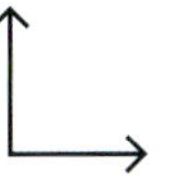

H90cm x
S50cm

Sun or
Part Shade

Moist Well
Drained Soil

Pollinators

Frost Hardy

Achillea

'TERRACOTTA'

A good-looking umbel, a member of the yarrow family, producing dense sprays of tiny red to orange florets June to September, loved by small pollinators such as hoverflies. The tall, upright stems raise the flowers above the feathery crown, but may need staking. Like many of these popular prairie-style plants, they offer handsome, sculptural seedheads that last well into winter. Cut them down to the ground before new growth emerges in spring. Plant in sun and a moist, well-drained soil.

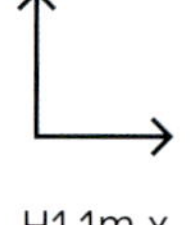

H1.1m x S40cm

Sun

Moist Well Drained Soil

Pollinators

Frost Hardy

PLANT WITH

Agastache
'Sandstone'

Upright spires of apricot-orange tubular flowers, July to October, with aromatic foliage. Ideal for a gravel garden and loved by bees. Cultivar 'Tango' is a richer orange. Grow in sun and moist, well-drained soil.

H30cm x S40cm

Agastache
'Blackadder'

Purple-flowered perennial giant hyssop forms attractive clumps of aromatic foliage and dense flower spikes July to October.

H90cm x S40cm

Miscanthus sinensis
'Morning Light'

Clump-forming silver grass, producing bronze flower plumes in October. Elegant and versatile, its blades have a cream stripe. Performs best in sun and moist, well-drained soil.

H1.2m x S1.2m

Calendula officinalis

'INDIAN PRINCE'

A favourite hardy annual companion plant for the veg patch, these multi-petalled marigolds flower June to October, are easy to grow, tolerant of most sites and soils, and will attract pollinators and beneficial pest-predators to your garden. The petals are edible, so make a whimsical addition to salads. Sow from seed March to May, then plant out in sun or part shade in moist, well-drained soil.

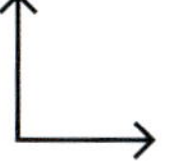

H60cm x S45cm

Sun or Part Shade

Moist Well Drained Soil

Pollinators

Frost Hardy

PLANT WITH

Tropaeolum majus
'Orange Troika'

Vigorous nasturtium with orange-red flowers and rounded, variegated leaves. Sow direct from March. Prefers full sun and moist, well-drained soil.

H30cm x S30cm

Centaurea cyanus

Vivid blue hardy annual loved by bees and other pollinators, flowering May to July or July to September – depending on when you sow them. Prefers a sunny spot in poor soil.

H70cm x S30cm

Dryopteris erythrosora

Semi-evergreen fern producing bronzed, triangular foliage. Plants form neat clumps ideal for a woodland look in a shady spot. Prefers moist, fertile soil improved with organic matter.

H1m x S1m

Dahlia
'DAVID HOWARD'

The elaborate orange pompom flowers of this popular 'decorative' dahlia are hard to beat for end-of-season drama, flowering July to October. The orange blooms stand out strikingly against the dark foliage, adding a brooding element ideal for a hot border or exotic planting scheme. Plant in a sheltered sunny spot and a moist, well-drained soil.

PLANT WITH

Canna
'Phasion'

Fiery, orange-flowered canna lily blooming June to September, with bold, paddle-shaped green, purple and gold striped leaves. The rhizomes won't survive a frost, so lift plants in late autumn, cut back the foliage and store in a frost-free place. Prefers sun and a moist, well-drained soil.

H1.6m x S50cm

Agastache
'Blackadder'

The bee-friendly, purple spikes of perennial giant hyssop emerge July to October, above clumps of aromatic foliage. Plant in a sunny spot in moist, well-drained soil.

H90cm x S40cm

Pennisetum setaceum
'Summer Samba'

Dark-leaved perennial fountain grass producing arching foliage and reddish-purple flower spikes from July to September. Plant in a sunny position in moist but well-drained soil. Plants are deciduous and not reliably hardy in a cold garden, so keep sheltered from frost.

H1.5m x S60cm

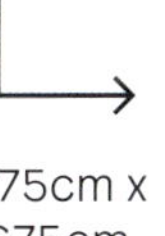

H75cm x S75cm

Sun

Moist Well Drained Soil

AGM Rated

Kniphofia

'TAWNY KING'

Red hot pokers offer an exciting explosion of colour, in a choice of hues including orange and yellow. This bright-orange cultivar produces its rocket-shaped blooms from July to October, emerging in succession from a clump of strappy foliage. They're ideal for a sunny spot or light shade, in a moist, well-drained soil.

PLANT WITH

Helenium
'Sahin's Early Flowerer'

A golden-yellow sneezeweed compared to redder, slightly taller 'Moerheim Beauty'. This striking perennial looks lovely in a prairie-style or hot border, flowering June to August above clumps of mid-green foliage. Plant in sun and a moist, well-drained soil.

H1m x S50cm

Agapanthus africanus

African lily flowering June to September, with starry, tubular blue trumpets held on stout stems above strappy foliage. Protect from frost in colder gardens, planting in a sunny sheltered spot in well-drained soil.

H60cm x S45cm

Cordyline australis
'Red Star'

Striking spiky-looking palm in deepest red, forming a tree in time. A popular species for seafront bedding schemes, but don't let that put you off! Plants are borderline hardy and will perish in a hard frost – so protect yours in winter. Plant in sun or light shade and moist, well-drained soil.

H3m x S2m

H1.25m x S1m

Sun or Part Shade

Moist Well Drained Soil

Frost Hardy

AGM Rated

Tithonia rotundifolia

'TORCH'

A flamboyant half-hardy annual that hails from Mexico, producing citrus-bright flowers from July to October. Get them started early from seed, sowing indoors March to April, and planting outdoors after all risk of frost has passed (towards end of May). They'll thrive in hot, sunny borders in a moist but well-drained soil.

PLANT WITH

Rudbeckia triloba
'Prairie Glow'

Exciting, if short-lived, perennial bearing two-toned flowers on wiry stems, July to October, in orange and yellow, with a buzzy brown nose cone. For sun or part shade and a moist, well-drained soil.

H1.2m x S70cm

Coreopsis grandiflora
'Early Sunrise'

Tightly petalled tickseed flowers, producing golden-yellow blooms from June to October. A short-lived perennial often grown from seed as an annual, sown March to May. Prefers sun or part shade in moist, well-drained soil.

H45cm x S45cm

Hordeum jubatum

Elegant squirrel-tail grass producing wafty, tufty flowers June to July on upright stems that bob in the breeze. A short-lived perennial commonly grown from seed as an annual, preferring sun and a moist, well-drained soil.

H50cm x S30cm

H1.5m x
S1m

Sun

Moist Well
Drained Soil

Pollinators

Rudbeckia hirta

'INDIAN SUMMER'

These large, golden daisies are sometimes called cone flowers or black-eyed Susans. They make a sumptuous addition to late-season borders and will shine from July to October. They'll thrive in sun, in moist but well-drained soil.

PLANT WITH

Dahlia
'David Howard'

Orange pompom flowers offer reliable hot-border drama, July to October, with dark, brooding foliage. Plant in a sheltered sunny spot and a moist, well-drained soil. Dahlia tubers resent a soggy clay soil in winter, so lift yours and store in a frost-free shed until spring, unless you can offer light, sandy soil and well-drained conditions.

H75cm x S75cm

Rudbeckia hirta
'Cherry Brandy'

Forming an attractive contrast for 'Indian Summer', this dazzling daisy has rich, carmine-red flowers on upright stems from July to November. Plants can be treated as half-hardy annuals but should overwinter in a sheltered, sunny spot. Sow undercover from February to April and plant out after all risk of frost.

H60cm x S30cm

Ricinus communis
'Carmencita'

The brooding, palmate foliage in reddish-purple will create an instant hot border effect, flowering July to September. The bright-red sputnik-shaped fruits contain toxic seeds, so handle with gloves when sowing.

H1.2m x S90cm

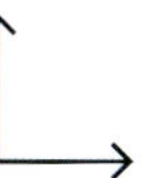

H60cm x
S50cm

Sun

Moist Well
Drained Soil

Pollinators

Frost Hardy

AGM Rated

Hamamelis x intermedia

'JELENA'

These fragrant orange flowers are an eye-catching winter treat, appearing on the bare stems of deciduous witch hazel from January to February. Shrubs form a vase-like shape ideal for anchoring a sunny, sheltered winter border – but give them space to achieve full height. The flowers are lightly scented too – so plant within sniffing distance. For sun or part shade and moist but well-drained soil.

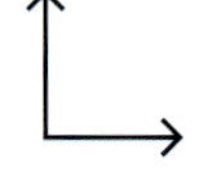

H4m x S4m

Sun or Part Shade

Moist Well Drained Soil

Frost Hardy

AGM Rated

PLANT WITH

Cornus sanguinea
'Anny's Winter Orange'

A vigorous, orange-stemmed dogwood that offers a bewildering range of seasonal treats – from its bright green buds to white June flowers, late summer berries, golden autumn foliage and, of course, those orange winter stems. Plant in sun or part shade in moist, well-drained soil.

H2.3m x S2.5m

Galanthus nivalis

Small, white, nodding flowers, January to March, with green markings and delicate pedicels. Grows best in sun or light shade around the base of deciduous trees and shrubs.

H10cm x S10cm

Nandina
'Firepower'

Small but dramatic evergreen shrub, or heavenly bamboo, whose leaves have a shapely, oriental look about them and flush scarlet in autumn. Starry white flowers in July. Plant in sun in moist, well-drained soil.

H45cm x S60cm

Cornus sanguinea

'MIDWINTER FIRE'

These spectacular two-toned winter stems dispel the seasonal gloom! Plant them in the sun for special effects that will turn neighbours' heads. In spring and summer bright green leaves emerge, and flowers in June. Autumn foliage will turn fiery orange-yellow before they fall. For best colour, cut back the stems hard in their second year of planting to encourage colourful, vigorous new growth in spring. Plant in sun or part shade and moist, well-drained soil.

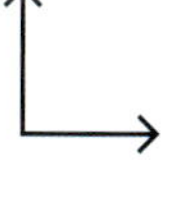

H2m x S2m

Sun or Part Shade

Moist Well Drained Soil

Pollinators

Frost Hardy

AGM Rated

PLANT WITH

Pyracantha
'Orange Glow'

Vigorous, spiny evergreen shrub, producing orange berries in autumn and winter. A good 'security plant' for the front garden! Best in sun or part shade, moist, well-drained soils and tolerant of pollution.

H2.5m x S2.5m

Cornus sericea
'Flaviramea'

Yellow-stemmed dogwood bearing small white flower clusters from May to June, attractive red autumn leaves and green-to-gold winter stems. Cool contrast for 'Midwinter Fire' in sun or light shade and moist, well-drained soil.

H1m x S1.5m

Fagus sylvatica
Atropurpurea Group

Copper beech is a deciduous tree, or hedge, bearing purple foliage in spring and summer, and catkins in May. Foliage turns copper in autumn and the younger leaves will remain on the plant for the winter. Plants can be clipped into topiary forms – a wonderful plant for sun.

Keep the hedge trimmed to less than H2m or H12m x S8m+ in 50 years

Build Your
Plant Matrix

+

+

+

+

+

+

+

+

Orange with...

Red	*Dahlia* 'David Howard' with red-leaved *Ricinus communis* 'Carmencita' (see page 36)
Pink	Nasturtiums (*Tropaeolum majus*) with *Astrantia major* 'Roma' (see page 59)
Purple	*Helenium* 'Sahin's Early Flowerer' with *Salvia nemorosa* 'Caradonna' (see page 102)
Blue	*Tulipa* 'Ballerina' with blue forget-me-nots, *Myosotis sylvatica* (see page 120)
Green	*Geum* 'Totally Tangerine' with ferns and hostas
Yellow	*Crocosmia* 'Emily Mackenzie' with *Rudbeckia fulgida* var. *deamii* (see page 206)
Orange	*Kniphofia* 'Tawny King' with *Achillea* 'Walther Funcke'
White	*Cornus sanguinea* 'Anny's Winter Orange' with *Galanthus nivalis* (see page 278)

White

WHITE is the colour of swans' wings, scudding clouds, moonlight and shooting stars. It is the traditional symbol of bridal purity and carries connotations of simplicity and elegance. It is the colour of blizzards blasting the arctic tundra; polar bears, igloos and snowmen. It is clean and clinical (doctors' lab coats, dental floss and dazzling smiles) and yet comfortingly suburban too: laundered sheets on a washing line and milk bottles on the doorstep. It is the white cliffs of Dover, and the tumbling tips of waves crashing onto the shore.

Because white light can be dispersed into all the colours of the spectrum, it's sometimes considered a non-colour. Yet for gardeners, white is very real. Flowers tend to be described as 'cool white' or 'creamy white' – a helpful distinction because the two don't always sit well together. There are green-whites (snowdrops, hydrangea 'Annabelle') and pink-whites (apple blossom) to consider too. The general advice is to avoid mixing cool and warm tones without some green in between.

Many white-themed gardens are inspired by Vita Sackville-West's famous one at Sissinghurst Castle in Kent. Restricting your colours to a monochromatic palette like this is a helpful exercise when you're learning about plants, as it forces you to focus, narrow your search, plan and edit your choices. And, while it's tempting to just fill up the shopping trolley with anything bearing white flowers each season, you quickly learn that blooms are fleeting and it's actually the many leaf colours, shapes and textures that play a more crucial role in creating a harmonious overall picture.

When you're drawing up your shopping list, remember that many traditionally pink-flowered plants also come in white. This is certainly true of foxgloves (*Digitalis*), echinacea, dicentra (*Lamprocapnos*), Japanese anemones, *Bergenia cordifolia* and *Clematis montana*. It's the same for some blue-flowering plants, such as forget-me-nots (*Myosotis*), borage, lavender, *Muscari armeniacum* and *Brunnera macrophylla*. The Latin word 'alba' or 'albiflora' in the species or cultivar name is a big clue. If you want to play with flower shapes and colour contrasts, you could do worse than team up pink echinaceas with white ones, pink foxgloves with white ones, and so on. They will always look lovely together.

One particular flower type to mention is the white umbel. There are so many of them it's mind boggling, and they look so similar even the experts sometimes struggle to tell them apart. Popular options are bishop's flower (*Ammi majus*), toothpick weed (*Ammi visnaga*), white laceflower (*Orlaya grandiflora*) and our native cow parsley (*Anthriscus sylvestris*), not to mention sweet cicely (*Myrrhis odorata*), wild carrot (*Daucus carota*), baltic cow parsley (*Cenolophium denudatum*) and wild parsnip (*Pastinaca sativa*). These lacey lovelies vary in height, spread and habit, but all create a light, billowing froth of summer flowers that contrasts perfectly with more solid forms such as tree trunks and rounded evergreens – pittosporum, euonymus, topiaried box and yew.

In the classic white garden, spring begins when the early bees start to notice the fragrant panicles of *Ribes sanguineum* 'White Icicle', underplanted with snowdrops and white daffodils, such as 'Thalia'. These are followed in rapid succession by tulips, including 'White Triumphator' and 'Spring Green', and white bluebells. White-flowered

Brunnera macrophylla or white forget-me-nots chime nicely too. Don't forget *Magnolia stellata*, whose open, starry flower shape is said to offer a bit of built-in design resistance to late frosts.

In late spring to early summer, a range of shade-tolerant white perennials starts to take off, including *Aquilegia* 'Nivea', *Astrantia* 'Large White' and bee-friendly foxgloves, *Digitalis purpurea* f. *albiflora*, which add height and vertical oomph.

Fragrant, white-flowered shrubs such as *Choisya ternata*, lilac (*Syringa vulgaris* 'Madame Lemoine') and sun-loving jasmine (*Trachelospermum jasminoides*) bring more white flowers in early summer. White *Buddleja davidii* and *Hydrangea macrophylla* cultivars bloom slightly later. Hide fences with decorous climbers such as vigorous *Clematis montana* var. *grandiflora*, followed by white rose 'Madame Alfred Carrière'.

Midsummer brings a treasure trove of white annuals and perennials into bloom: lavender, phlox, cosmos, gaura (*Oenothera*), scabious, honesty (*Lunaria annua*), thalictrum, astilbe, sweet peas, nicotiana, centaurea, nigella, hollyhocks. You really are spoiled for choice, which is where the editing comes in, mentioned before. And we've not even mentioned the many different white daisies – the Shasta daisy (*Leucanthemum* x *superbum*) being a cottage-garden favourite, but also Mexican fleabane, *Erigeron karvinskianus*, which is an enthusiastic self-sower but very good at creating a low-level froth at the path edge, or mounds of pink to white daisies around the garden.

Autumn whites include the Michaelmas daisy, *Symphyotrichum novi-belgii* 'White Ladies', August to October, and also the fabulous Japanese anemone, *Anemone* x *hybrida* 'Honorine Jobert'. People say the latter is less vigorous (i.e. less invasive) than the pink species, but in rich clay soil this pretty perennial doesn't need much encouragement. It's a lovely problem to have, and worth it for those elegant late flowers on tall, wiry stems.

Stepping into winter, *Cyclamen hederifolium f. albiflorum* will cheer up a bare border with its white flowers and marbelled leaves from October into November. Late-flowering evergreen *Clematis cirrhosa* 'Jingle Bells' and *C. urophylla* 'Winter Beauty' will bring bell-shaped flowers from December to February, followed by attractive seedheads. Other cold-season treats include the winter honeysuckles – *Lonicera fragrantissima* and *L.* x *purpusii* 'Winter Beauty' – both with fragrant, tubular flowers on bare stems. *Helleborus niger*, the Christmas rose, will bring your white border safely into the new year, and then it's snowdrop time again. Plant a few white winter pansies as a pick me up, among glossy evergreens and vibrant dogwoods.

Magnolia stellata

The starry white flowers of this statuesque small tree make it a popular choice for early spring flowers, March to April. Easy to care for, mulch in spring with well-rotted manure. Magnolias famously succumb to a late frost, which will brown the buds and flowers, so plant in a sheltered spot, in sun and moist, well-drained soil.

PLANT WITH

Leucojum
'Gravetye Giant'

Resembling a giant snowdrop, summer snowflake flowers bloom taller and later, from March to May, forming clumps of grass-like, strappy green foliage. Plant the bulbs in autumn 10cm deep in sun or part shade and a moist, well-drained soil.

H50cm x S10cm

Primula vulgaris

Delicate native primrose has creamy yellow flowers from March to May, on short stems above rosettes of crimped green foliage. Ideal in a shady border under deciduous trees, or among ferns and other shade lovers in moist, well-drained soil.

H20cm x S35cm

Taxus baccata

The dark green, needle-like leaves of yew are easy to keep trimmed into topiary mounds, pyramids or small hedges. They're an excellent foil for any white flower, adding a formal note to more relaxed planting schemes. Plant in light shade in moist, well-drained soil.

H12m x S8m+ in 50 years

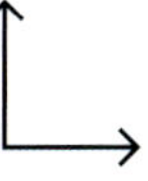

H3m x
S4m

Sun or
Part Shade

Moist Well
Drained Soil

Narcissus
'THALIA'

With its multi-headed flower stems, swept-back outer petals and long white trumpets, this demure daffodil is perfect for a white border, flowering March to April. Clumps naturalize well in beds and borders, but the bulbs can also be planted in grass, in autumn, about 15cm deep. Best in sun or light shade in moist but well-drained soil.

PLANT WITH

Brunnera macrophylla
'Mr Morse'

Brunnera with heart-shaped, marbled leaves very much like 'Jack Frost', only with delicate sprays of white flowers, produced March to April. Plants prefer a part-shady position in moist but well-drained soil.

H45cm x S45cm

Tulipa
'Doll's Minuet'

Vibrant, lily-flowered magenta-red tulip flushed with green, flowering April to May. A colourful, shapely contrast for white tulip 'Spring Green'. Plant the bulbs in autumn in sun or part shade and a moist, well-drained soil.

H55cm x S15cm

Dryopteris affinis
'Cristata'

Native woodland fern producing clumps of lush, semi-evergreen foliage. Good for a woodland setting or under a tall shrub whose canopy has been raised by removing lower branches. Prefers part shade and a moist, well-drained soil.

H90cm x S90cm

H40cm x
S15cm

Sun or
Part Shade

Moist Well
Drained Soil

Pollinators

Frost Hardy

Exochorda x *macrantha*

'THE BRIDE'

H2m x S3m

This dramatic deciduous shrub is densely decked with white flowers from March to May, flowering on bare, arching branches. Its fluttering spring blossom creates carpets of confetti, followed by summer leaves that turn gold in autumn, making this a sensational scene-stealer for a sunny, or part shady, corner of the garden in moist but well-drained soil.

Sun or Part Shade

Moist Well Drained Soil

Pollinators

Frost Hardy

AGM Rated

PLANT WITH

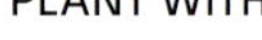

Tulipa
'Spring Green'

Elegant, goblet-shaped white flowers on upright stems. Plant the bulbs in autumn for flowers April to May. Tulips prefer a sunny spot and soil that doesn't become waterlogged. Dig in horticultural grit to improve a heavy-clay site.

H60cm x S10cm

Clematis alpina

Nodding, purple, bell-shaped flowers smother this spring-flowering clematis from April to May, followed by fluffy seedheads. Free-flowering cultivar 'Helsingborg' is shown here. Plants are deciduous, so will lose their leaves over winter, and prefer sun or light shade in moist, well-drained soil.

H3m x S1.5m

Betula utilis var. *jacquemontii*

This popular Himalayan birch is a garden-designer favourite, thanks to its bold white trunks, which can be gently washed with water to brighten them up. Gold catkins and colourful autumn leaves add to the seasonal show. Multi-stem trees create more impact in a medium-sized garden – but mind you have sufficient space.

H12m x S8m

Viburnum opulus

'ROSEUM'

The snowball tree is a popular choice for any garden, with its rounded pompom flowers from May to June. Buds open to fist-sized flower clusters that start off green-white and mature to a pinker hue. It's quite the scene-stealer in spring, and the leaves offer good autumn colour too. Plant in sun and moist, well-drained soil.

H5m x S5m

Sun

Moist Well Drained Soil

Pollinators

Frost Hardy

AGM Rated

PLANT WITH

Digitalis purpurea f. *albiflora*

Handsome biennial foxglove producing a rosette of foliage in its first year and flowering the second. Plants are happy to self-sow, ensuring white foxgloves every year. Best in light shade and moist, humus-rich soil.

H1.5m x S60cm

Primula veris

Native cowslip bears yellow flower clusters on upright stems, like a tiny standard lamp with a fancy shade, April to June. Plants will steadily increase and naturalize in a wildflower border, in sun or light shade.

H25cm x S25cm

Polystichum setiferum

Evergreen soft shield fern whose new croziers unfurl from knobbly knuckles in spring. Cut back the old foliage in January. Plants prefer a shady position in moist, well-drained, humus-rich soil.

H1.2m x S90cm

Tulipa
'WHITE TRIUMPHATOR'

These elegant, goblet-shaped, satinesque flowers are held on upright stems that look especially charming planted en masse. Plant the bulbs in autumn and they'll flower the following April to May. Tulips prefer a sunny spot and soil that doesn't become too wet and waterlogged. They're hardy but may not reliably return the following year.

PLANT WITH

Erythronium californicum
'White Beauty'

Pretty dog's-tooth violet, with swept-back white petals that look a bit like lilies on slender stems above clumps of marbled foliage. Lovely in a shady woodland border, in moist, well-drained soil.

H30cm x S30cm

Fritillaria meleagris

Native snake's-head fritillary producing unusual nodding flowers with a chessboard pattern in maroon and white-pink. Ideal for growing in a wildflower border or long grass; a bold companion for white tulips, too. Plant in sun or part shade and moist, well-drained soil.

H30cm x S8cm

Euonymus fortunei
'Emerald Gaiety'

Handsome shrub with green and white variegated evergreen foliage. Very useful as a backdrop plant in a white-themed garden. Leaves can take on a pink blush as temperatures drop in winter. Plant in sun or light shade, in moist well-drained soil.

H1m x S1.5m

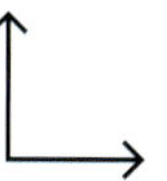

H60cm x S10cm

Sun

Moist Well Drained Soil

Frost Hardy

AGM Rated

Digitalis purpurea f. *albiflora*

White foxgloves are such versatile plants for any planting scheme – very much at home in a cottage garden but making a chic addition to shady, modern, urban gardens too. These handsome biennial plants will produce a rosette of foliage in their first year and flower the second; if you shake the seedheads around the garden, they'll self-sow and ensure a succession of foxglove babies every spring. Best in light shade and moist, humus-rich soil.

PLANT WITH

Anthriscus sylvestris
'Ravenswing'

Native cow parsley is one of many white-flowered umbels; this cultivar, however, has purple stems, which makes it a bit fancy. Plants are biennial or short-lived perennials, flowering in their second year, May to July. Plant en masse for billowing clouds of bloom ideal for a naturalistic woodland area in sun or light shade and moist, well-drained soil.

H1m x S30cm

Lupinus
'Masterpiece'

Fabulous purple perennial producing chubby, skyscraper flowers in June, above clumps of attractive palmate leaves. Deadhead for a second flush in late summer but keep an eye out for aphids: blast them off with the hose.

H90cm x S60cm

Hosta
'Patriot'

Neatly ribbed, medium-sized foliage plant with green leaves and white edges. Protect from molluscs using garlic spray (apply fresh after rain) or if in containers, copper bands circling the rim. Lavender-blue flowers July to August. Plant in part shade and moist, well-drained soil.

H75cm x S1m

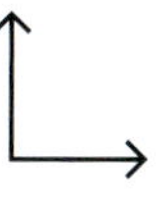

H1.5m x
S60cm

Part Shade or
Full Shade

Moist Well
Drained Soil

Pollinators

Frost Hardy

Trachelospermum jasminoides

A fragrant-flowered, sun-loving evergreen climber with glossy green foliage and small, starry, white flowers from June to August. Plant against an obelisk or trellis in a sheltered sunny spot, near a door where you'll notice the fragrance. It's only borderline hardy, so protect from frost in cold gardens.

PLANT WITH

Leucanthemum vulgare

Statuesque, native perennial white daisy with yellow nose, flowering May to July. Popular with pollinators, including bees and butterflies. Plant in sun and a moist, well-drained soil.

H90cm x S60cm

Verbena bonariensis

Pink or light-purple-flowered, slender-stemmed perennial whose tightly packed flower clusters bloom June to October. Beautiful and useful, plant in sun and a moist, well-drained soil.

H2m x S45cm

Fatsia japonica

Lustrous-leaved hardy shrub forming a glossy evergreen backdrop, with a slightly exotic look. Produces little white sputnik flowers and black fruit in autumn – can reach glorious dimensions in a sunny or shady corner in moist, well-drained soil.

H4m x S4m.

H9m x
S6m

Sun

Moist Well
Drained Soil

Pollinators

AGM Rated

Philadelphus

Mock orange has deliciously fragrant, creamy-white double flowers from April to June, guaranteed to turn the heads of passers-by, wondering where that fabulous smell is coming from. Popular cultivars include 'Belle Etoile', 'Manteau d'Hermine' and 'Snowbelle', the latter with creamy double flowers June to July. Plant in a moist but well-drained soil.

PLANT WITH

Hosta sieboldiana var. *elegans*

Large-leaved hosta with ribbed, puckered foliage and pretty clusters of pale lilac, tubular flowers July to August. Ideal for a woodland look under the light leafy canopy of a small tree or tall shrub, with ferns. Plants prefer a shady spot in moist, fertile soil.

H1m x S1.2m

Rosa
'The Fairy'

Repeat-flowering shrub rose, producing fragrant, blowsy pink double blooms from July to September. A classic cottage-garden bloom. Plant in sun or part shade in a rich, moist but well-drained soil.

H1m x S1m

Pittosporum tenuifolium
'Silver Queen'

Glossy evergreen shrub with white-edged, curvy leaves. Casting light, dappled shade, you can plant all manner of white shrubs and perennials in its shelter. Raise the canopy by removing lower branches for a more tree-like habit. Grow in sun, sheltered from frost, in moist, well-drained soil.

H4m x S4m

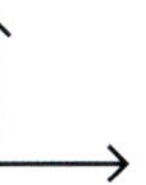

H3m x
S2.5m

Sun or
Part Shade

Moist Well
Drained Soil

Frost Hardy

AGM Rated

Orlaya grandiflora

One of many attractive white umbels, orlaya has intricate, lace-cap-type heads combining small flowers surrounded by larger petals June to October. Foliage is fine and feathery. For best results, plant this hardy annual in sun and moist, well-drained soil.

H60cm x S60cm

Sun

Moist Well Drained Soil

Pollinators

Frost Hardy

AGM Rated

PLANT WITH

Agapanthus africanus 'Albus'

White-flowered African lily, producing dramatic spheres of tubular flowers on tall stems, July to September, like miniature white firework explosions. Plant in sun and a moist, well-drained soil.

H60cm x S45cm

Cosmos bipinnatus

With their open flowers and fern-like foliage, cosmos go with everything. Available in a range of pinks and white, such as 'Purity', grow them from seed March to May for planting out after all risk of frost, in a sunny position in moist, well-drained soil.

H1.2m x S45cm

Stipa tenuissima

Wispy-looking, clump-forming grass whose semi-evergreen blades emerge green then fade to buff, with feathery flowers June to October. A good self-sower that goes with everything, it's ideal for introducing movement into a planting scheme. Best in sun and moist, well-drained soil.

H60cm x S30cm

Oenothera lindheimeri

'WHIRLING BUTTERFLIES'

Previously known as *Gaura*, this delicate, fluttery-flowered perennial adds the gentlest touch of pink to a white border. The airy flower stems have a graceful habit, May to September, emerging among leafy stems. Plants are borderline hardy but drought-tolerant once established. Plant in a sunny spot in moist, well-drained soil.

H75cm x S85cm

Sun

Moist Well Drained Soil

Pollinators

Frost Hardy

PLANT WITH

Erigeron karvinskianus

This charming perennial is a successful self-sower that will spread itself around the garden, creating mounds of pink and white daisies, June to October. Loved by butterflies. Plant in sun and moist, well-drained soil.

H30cm x S1m

Echinacea purpurea

Iconic, pink-flowered daisy whose drooping petals and fuzzy orange-brown nose cone are loved by bees and butterflies alike, June to September. Upright stems last well across the season; the flowers are followed by seedheads attractive to small garden birds. Cultivar 'Magnus' is shown here. Plant in sun and a moist, well-drained soil.

H1.5m x S45cm

Miscanthus sinensis 'Morning Light'

Perennial clump-forming grass with arching, silvery leaf blades with a cream margin and feathery flower tassels in October. An elegant option for adding movement to a border. For sun and a moist, well drained soil.

H1.2m x S1.2m

Phlox paniculata

'EARLY WHITE'

Stunning *Phlox paniculata* is a favourite perennial for mixed herbaceous borders, producing large, blowsy flowerheads July to September with a honey-like fragrance. There are lots of attractive white cultivars to choose from, including 'David', 'Mount Fuji' and 'Rembrandt' (H90cm), as well as low-growing *P. subulata* 'Snowflake' (H15cm). Plants prefer sun or part shade and a moist, well-drained soil.

PLANT WITH

Leucanthemum vulgare

Robust, native perennial ox-eye daisy that teams well with all manner of tall perennial flowers and grasses. Popular with bees and butterflies, it prefers sun and a moist, well-drained soil.

H90cm x S60cm

Verbena bonariensis

Pink or light-purple-flowered perennial bearing dense flower clusters from June to October, on tall wiry stems. Use it as a light screening plant, adding accent colour in a white border. Plant in sun and a moist, well-drained soil.

H2m x S45cm

Heuchera
'Green Spice'

Perennial groundcover plant with scalloped, green-edged leaves traced with red margins. Gorgeous colouring that looks good with other red and green heucheras. Sprays of small white flowers emerge June and July on tall stems. Ideally plant in sun and a moist, well-drained soil.

H35cm x S35cm

H45cm x
S35cm

Sun or
Part Shade

Moist Well
Drained Soil

Pollinators

Frost Hardy

Erigeron karvinskianus

Vigorous Mexican fleabane is a successful self-sower that will spread itself around the garden and into all your paving cracks in no time, creating pretty, trailing mounds of pink and white daisies, June to October. This generous plant can be used to create a billowing fringe at the path edge, or a tumbling cascade of flowers from a wall. Loved by butterflies. Plant in sun and moist, well-drained soil, though it needs very little encouragement.

PLANT WITH

Echinacea purpurea
'White Swan'

Eye-catching white cousin of the more familiar pink echinacea. The petals looks slightly droopy, like a badminton shuttlecock, with a fuzzy-looking gold nose, flowering June to September. A good choice for pollinators. Plant in sun and moist, well-drained soil.

H60cm x S45cm

Campanula portenschlagiana

Vigorous alpine perennial with dinky purple flowers on mats of mid-green, ivy-like foliage. A good spreader that looks lovely tumbling over stone walls. Plant in sun or part shade in light, well-drained soil.

H15cm x 50cm

Festuca glauca
'Elijah Blue'

Silver-blue fescue forming rounded clumps, with buff-coloured flowers June to July. Looks good with silver foliage plants and anything with white flowers. For sun or part shade and moist, well-drained soil.

H30cm x S25cm

H30cm x S1m

Sun

Moist Well Drained Soil

Pollinators

Frost Hardy

AGM Rated

Rosa

'FÉLICITÉ PERPÉTUE'

H6m x S4m

A fragrant rambling rose, the French name means 'Happy Ever After'. The large clusters of small, creamy-white flowers are flushed pink at first, emerging from deep pink buds and flowering July to September. Perfect for a large pergola or tricky shady area, plants prefer sun or light shade and moist, well-drained soil.

Sun or Part Shade

Moist Well Drained Soil

Pollinators

Frost Hardy

AGM Rated

PLANT WITH

Verbascum chaixii
'Album'

Upright mullein producing white flower stems June to August with a boss of pink filaments at the centre. Loved by pollinators, it teams well with both matching white and contrasting pink flowers. Plant in sun and moist, well-drained, slightly alkaline soil. Can fall victim to mullein moth – pick the caterpillars off by hand and pop them on the bird table.

H90cm x S45cm

Nepeta racemosa
'Walker's Low'

Perennial catmint bearing purple flowers on slender stems from June to September. Plants form generous clumps of leaf and lax stems that tumble attractively over path edges for a cottage-garden look. Plant in sun or part shade in moist, well-drained soil.

H60cm x S60cm

Stipa tenuissima

Clump-forming grass whose green leaf blades fade to blonde, with feathery flowers June to October. A reliable self-sower that goes with everything. Plant in sun and moist, well-drained soil.

H60cm x S30cm

Hydrangea arborescens

'ANNABELLE'

Popular, distinctive hydrangea carrying enormous round white flowerheads from July to September. The flowers can be heavy and droopy; stems of newer cultivar 'Strong Annabelle' are more upright. Good autumn colour and the flowers look lovely cut and dried for the vase. Grow in part shade and keep watered in drought.

PLANT WITH

Hosta sieboldiana* var. *elegans

Elegant perennial hosta with lime-green foliage and palest lilac flowers July to August, which harmonize well with a white planting scheme. The large, ribbed leaves form rounded heart shapes, adding to the appeal. Protect from slugs in early spring.

H1m x S1.2m

Eupatorium maculatum

Joe Pye weed is a statuesque, back-of-border, butterfly-friendly perennial bearing fluffy pink flower umbels July to September on upright purple stems. A dramatic screening choice ideal for a prairie planting scheme. Plant in sun or part shade, in moist, well-drained soil.

H2m x S2m

Hakonechloa macra

Mid-green, shaggy-looking Japanese grass with relaxed leaf blades forming a cascade at the border edge. Will form an attractive 'hula' skirt around the ankles of a small tree. Full sun or light shade and a moist, well-drained soil.

H35cm x S40cm

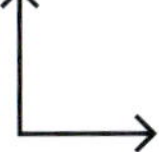

H2.5m x S2.5m

Part Shade

Moist Well Drained Soil

Frost Hardy

AGM Rated

Cosmos bipinnatus

'PURITY'

These soft, silky, half-hardy annual daisies are easy to grow from seed and literally go with everything. Sow March to May and plant out after all risk of frost has passed for flowers June to October that seem to float above the feathery foliage. Keep deadheading for flowers to the frosts. Best in sun and moist, well-drained soil.

PLANT WITH

Ammi majus

Bishop's weed is a native annual producing lacy white umbel flowers June to September. Sow March to May and plant out after all risk of frost for airy plants whose flowers seem to float above the foliage. Prefers sun or light shade in a moist, well-drained soil.

H1m x S50cm

Verbena bonariensis

Pink-flowered perennial bearing dense flower clusters from June to October, on tall wiry stems. Use it as a light screening plant, adding accent colour in a white border. Plant in sun and a moist, well-drained soil.

H2m x S45cm

Foeniculum vulgare

Herbaceous perennial producing clouds of light, feathery foliage and yellow umbel flowers July to August, followed by sculptural seedheads. Plants are large and architectural, preferring full sun and moist, well-drained soil.

H1.8m x S45cm

H1.2m x S60cm

Sun

Moist Well Drained Soil

Pollinators

Anemone x hybrida

'HONORINE JOBERT'

Good-looking Japanese anemone with delicate white flowers held on tall, upright, wiry stems, from August to October. Plants form large clumps of shapely green leaves held on stiff green stems, creating presence and offering ground cover throughout the season. They can be a bit thuggish in rich soil but are easy to pull out if they overstep. Plant in sun or light shade and moist, well-drained soil.

PLANT WITH

Selinum wallichianum

Himalayan perennial bearing lacy white umbels July to September on upright stems, above fern-like foliage. The flowers are followed by attractive seedheads with wildlife appeal. Plant in sun or part shade in a moist, well-drained soil.

H1.2m x 90cm

Bistorta amplexicaulis 'Firetail'

Red bistort has knobbly red flower spikes July to October, held on slender upright stems. Loved by beneficial insects and pollinators, it's a good clump-forming groundcover plant for a sunny or tricky part-shady spot in moist, well-drained soil.

H1.2m x 1.2m

Calamagrostis x acutiflora 'Karl Foerster'

Versatile upright perennial grass producing green, fading to buff, flowers June to August. Ideal companion for taller perennials or used as a screen, adding height and movement to any planting scheme. A must-have for sun or part shade and moist, well-drained soil.

H1.8m x S60cm

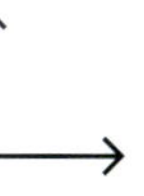

H1.2m x
1.2m

Sun or
Part Shade

Moist Well
Drained Soil

Pollinators

Frost Hardy

AGM Rated

Clematis cirrhosa

'JINGLE BELLS'

A nodding, bell-flowered evergreen climber that will transform fences and trellis panels with a mass of white flowers, December to February. Grow it in sight of a window, from where you can enjoy the flowers from indoors; they have a delicate scent that makes it worth venturing outside on a sunny day. Plants prefer sun or part shade in moist but well-drained soil.

H5m x S2m

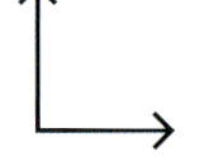

Sun or Part Shade

Moist Well Drained Soil

Pollinators

Frost Hardy

PLANT WITH

Galanthus nivalis

Demure, nodding, bell-shaped white flowers January to March, hanging from a slender pedicel, with green markings on the inner tepals. Plant in sun or part shade in moist well-drained soil.

H10cm x S10cm

Clematis cirrhosa 'Freckles'

Evergreen climber with nodding, bell-like cream flowers bearing pink flecks, December to February. Lovely partner for other winter clematis and attractive to early solitary bees on the wing. Plant in sun or part shade and a moist, well-drained soil.

H4m x S1.5m

Ilex aquifolium 'Argentea Marginata'

Female holly with bold silver edges to its prickly green leaves and spring flowers. A lovely evergreen for winter interest. Plant in sun or part shade in a moist, well-drained soil.

H15m x S4m

Lonicera fragrantissima

Handsome, semi-evergreen winter-flowering honeysuckle bearing dainty, white flowers like ruffled tutus, each with long white filaments holding anthers like golden ballet slippers, January to March. These dinky flowers are held on bare stems and really pump out the fragrance on a sunny day. Best trained against a sunny wall in moist, well-drained soil.

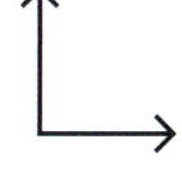

H2m x S3m

Sun

Moist Well Drained Soil

Pollinators

Frost Hardy

PLANT WITH

Helleborus niger

Nodding Christmas rose blooms January to February, its cup-shaped flowers flanked by thick, leathery palmate leaves. Plant in light shade and a moist, humus-rich soil.

H30cm x S45cm

Edgeworthia chrysantha

Deciduous shrub whose dense yellow flower clusters are produced on bare stems, February to April. Plant in a sunny, sheltered position and mulch well in spring.

H1.5m x S1.5m

Buxus sempervirens

A lovely, slow-growing, tactile evergreen for topiary effects and low parterre-style hedging, which alas often falls victim to box caterpillars and blight. Alternatives such as *Ilex crenata* and *Euonymus* 'Green Spire' lack the charm but offer a good small-leaved alternative for clipping to shape.

H5m x S5m if unclipped

Helleborus niger

Lovely, demure-looking Christmas rose, with downcast flowers, January to February. Foliage is palmate, thick and leathery, offering dark good looks that help the white flowers stand out. Plant in light shade and a moist, humus-rich soil.

PLANT WITH

Galanthus nivalis

Nodding, bell-shaped white flowers January to March, with green markings on the inner tepals. Lovely grown in a naturalized drift. Plant in sun or part shade in moist, well-drained soil.

H10cm x S10cm

Narcissus 'Golden Dawn'

Elegant daffodil with yellow petals and orange trumpet, flowering March to April. Loved by bees and other spring pollinators. Plant in sun in a moist, well-drained soil.

H40cm x S10cm

Ophiopogon planiscapus 'Nigrescens'

A dark-leaved perennial evergreen, black mondo grass makes a dramatic contrast for white-flowered companions. Grow it at the front of a border from where you can see it.

H20cm x S30cm

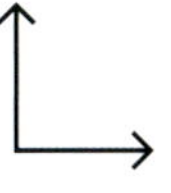

H30cm x
S45cm

Part Shade

Moist Well
Drained Soil

Pollinators

Frost Hardy

Galanthus nivalis

Iconic winter plant with slender, bell-shaped white flowers January to March, hanging from a slender pedicel, with green markings on the inner tepals. Buy and plant them in the green after flowering, as the dry bulbs are not always reliable into flower. Excellent naturalized in drifts and small informal groups under a dappled leaf canopy or deciduous trees and large shrubs. Utterly lovely and hugely collectable! Plant in sun or part shade in moist, well-drained soil.

PLANT WITH

Helleborus niger

The nodding white flowers of the Christmas rose bloom January to February, among thick, leathery palmate leaves. A bright 'pick-me-up' flower blooming just when you need it, in the dead of winter. Plant in light shade and a moist, humus-rich soil.

H30cm x S45cm

Crocus tomassinianus

Silvery-lilac-flowered 'tommys' are easy to naturalize in lawn grass, or in borders under deciduous trees before the summer foliage blocks out their sunlight. Plant the bulbs 10cm deep in a sunny position or light shade, in moist, well-drained soil.

H10cm x S3cm

Dryopteris affinis 'Cristata'

Native fern producing clumps of semi-evergreen foliage. Good for a woodland setting or under the raised canopy of a tall shrub. Prefers part shade and a moist, well-drained soil.

H90cm x S90cm

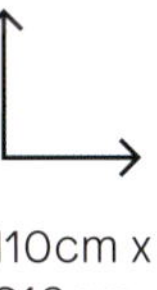

H10cm x
S10cm

Sun or
Part Shade

Moist Well
Drained Soil

Pollinators

Frost Hardy

AGM Rated

Build Your
Plant Matrix

White with...

+

+

+

+

+

+

+

+

Red	*Helleborus niger* with *Cornus alba* 'Sibirica' (see page 16)
Pink	*Echinacea purpurea* 'White Swan' with *Echinacea purpurea* 'Magnus' (see page 66)
Purple	*Leucanthemum vulgare* with *Salvia nemorosa* 'Caradonna' (see page 102)
Blue	*Cosmos bipinnatus* 'Purity' with blue delphiniums (see page 128)
Green	*Hydrangea* 'Annabelle' with aquilegia 'Green Apples' (see page 156)
Yellow	*Galanthus nivalis* with winter aconites, *Eranthis hyemalis* (see page 180)
Orange	*Ammi visnaga* with *Dahlia* 'David Howard' (see page 234)
White	White foxglove *Digitalis purpurea* f. *albiflora* with cow parsley *Anthriscus sylvestris* 'Ravenswing'

Picture Credits

a= above, b=below, l=left, r=right, c=centre

16l Sarah Cuttle © BBC Gardeners' World Magazine; 16c, 16r, 17 Jason Ingram © BBC Gardeners' World Magazine; 18l Sarah Cuttle © BBC Gardeners' World Magazine; 18c Jason Ingram © BBC Gardeners' World Magazine; 18r, 19 Sarah Cuttle © BBC Gardeners' World Magazine; 20l VladimirShnip; 20c Jason Ingram © BBC Gardeners' World Magazine; 20r Sarah Cuttle © BBC Gardeners' World Magazine; 21 Torie Chugg © BBC Gardeners' World Magazine; 22l, 22c, 22r Jason Ingram © BBC Gardeners' World Magazine; 23 Sarah Cuttle © BBC Gardeners' World Magazine; 24l VladimirShnip; 24c, 24r Jason Ingram © BBC Gardeners' World Magazine; 25 Sergey V Kalyakin; 26l Torie Chugg © BBC Gardeners' World Magazine; 26c Jason Ingram © BBC Gardeners' World Magazine; 26r Helen Pitt; 27, 28a Jason Ingram © BBC Gardeners' World Magazine; 28bl Sergey V Kalyakin; 28bc Jason Ingram © BBC Gardeners' World Magazine; 28br Lois GoBe; 29a, 29bl, 29bc Jason Ingram © BBC Gardeners' World Magazine; 29br Summer 1810; 30l, 30c, 30r Jason Ingram © BBC Gardeners' World Magazine; 31, 32a Sarah Cuttle © BBC Gardeners' World Magazine; 32b Svetliy; 32bc, Sarah Cuttle © BBC Gardeners' World Magazine; 32br JohnatAPW; 33a, 33bl, 33bc, 33br Jason Ingram © BBC Gardeners' World Magazine; 34l Alex Manders; 34c Jason Ingram © BBC Gardeners' World Magazine; 34r Sarah Cuttle © BBC Gardeners' World Magazine; 35 Torie Chugg © BBC Gardeners' World Magazine; 36bl Jason Ingram © BBC Gardeners' World Magazine; 36bc Sarah Cuttle © BBC Gardeners' World Magazine; 36br JohnatAPW; 37 3DF mediaStudio; 38l, 38c Sarah Cuttle © BBC Gardeners' World Magazine; 38r Jason Ingram © BBC Gardeners' World Magazine; 39 Sarah Cuttle © BBC Gardeners' World Magazine; 40a Tim Sandall © BBC Gardeners' World Magazine; 40bl Sarah Cuttle © BBC Gardeners' World Magazine; 40bc Marie Shark; 40br, 41a Sarah Cuttle © BBC Gardeners' World Magazine; 41bl JohnatAPW; 41bc, 41br Jason Ingram © BBC Gardeners' World Magazine; 42a, 42bl, 42bc, 42br Sarah Cuttle © BBC Gardeners' World Magazine; 43a N.Stertz; 43bl Jason Ingram © BBC Gardeners' World Magazine; 43bc Sarah Cuttle © BBC Gardeners' World Magazine; 43br Jason Ingram © BBC Gardeners' World Magazine; 48l Sarah Cuttle © BBC Gardeners' World Magazine; 48c, 48r, 49, 50l Jason Ingram © BBC Gardeners' World Magazine; 50c Sarah Cuttle © BBC Gardeners' World Magazine; 50r AnnaNel; 51, 52l, 52c, 52r Jason Ingram © BBC Gardeners' World Magazine; 53 Sarah Cuttle © BBC Gardeners' World Magazine; 54a Edita Medeina; 54bl, 54bc Jason Ingram © BBC Gardeners' World Magazine; 54br Sarah Cuttle © BBC Gardeners' World Magazine; 55a Irina Kvyatkovskaya; 55bl Sarah Cuttle © BBC Gardeners' World Magazine; 55bc Jason Ingram © BBC Gardeners' World Magazine; 55br, 56l Torie Chugg © BBC Gardeners' World Magazine; 56c Sarah Cuttle © BBC Gardeners' World Magazine; 56r, 57, 58a Jason Ingram © BBC Gardeners' World Magazine; 58bl Torie Chugg © BBC Gardeners' World Magazine; 58bc Jason Ingram © BBC Gardeners' World Magazine; 58br Sarah Cuttle © BBC Gardeners' World Magazine; 59a Jason Ingram © BBC Gardeners' World Magazine; 59bl, 59bc Sarah Cuttle © BBC Gardeners' World Magazine; 59br Jason Ingram © BBC Gardeners' World Magazine; 60l Torie Chugg © BBC Gardeners' World Magazine; 60c Jason Ingram © BBC Gardeners' World Magazine; 60r Sarah Cuttle © BBC Gardeners' World Magazine; 61–64 Jason Ingram © BBC Gardeners' World Magazine; 65, 66l Torie Chugg © BBC Gardeners' World Magazine; 66c Sarah Cuttle © BBC Gardeners' World Magazine; 66r, 67 Jason Ingram © BBC Gardeners' World Magazine; 68l Tim Sandall © BBC Gardeners' World Magazine; 68c, 68r, 69, 70l, 70c Jason Ingram © BBC Gardeners' World Magazine; 70r P Tomlins/Alamy Stock Photo; 71 Jason Ingram © BBC Gardeners' World Magazine; 72l Torie Chugg © BBC Gardeners' World Magazine; 72c © Liz Potter; 72r Jason Ingram © BBC Gardeners' World Magazine; 73 Torie Chugg © BBC Gardeners' World Magazine; 74a Sarah Cuttle © BBC Gardeners' World Magazine; 74bl Jason Ingram © BBC Gardeners' World Magazine; 74bc Tim Sandall © BBC Gardeners' World Magazine; 74br Neil Hepworth © BBC Gardeners' World Magazine; 75a Jason Ingram © BBC Gardeners' World Magazine; 75bl Sarah Cuttle © BBC Gardeners' World Magazine; 75bc Jason Ingram © BBC Gardeners' World Magazine; 75br Sarah Cuttle © BBC Gardeners' World Magazine; 80l, 80c Jason Ingram © BBC Gardeners' World Magazine; 80r, 81, 82l Sarah Cuttle © BBC Gardeners' World Magazine; 82c, 82r Jason Ingram © BBC Gardeners' World Magazine; 83 Orest lyzhechka; 84a Sarah Cuttle © BBC Gardeners' World Magazine; 84bl Jason Ingram © BBC Gardeners' World Magazine; 84bc Tim Sandall © BBC Gardeners' World Magazine; 84br, 85a, 85bl, 85bc Jason Ingram © BBC Gardeners' World Magazine; 85br Sarah Cuttle © BBC Gardeners' World Magazine; 86l, 86c Jason Ingram © BBC Gardeners' World Magazine; 86r, 87 Sarah Cuttle © BBC Gardeners' World Magazine; 88l Torie Chugg © BBC Gardeners' World Magazine; 88c Jason Ingram © BBC Gardeners' World Magazine; 88r © Liz Potter; 89, 90l Jason Ingram © BBC Gardeners' World Magazine; 90c Torie Chugg © BBC Gardeners' World Magazine; 90r, 91 Jason Ingram © BBC Gardeners' World Magazine; 92l Sarah Cuttle © BBC Gardeners' World Magazine; 92c, 92r, 93–95, 96a 96bl Jason Ingram © BBC Gardeners' World Magazine; 96bc Neil Hepworth © BBC Gardeners' World Magazine; 96br Sarah Cuttle © BBC Gardeners' World Magazine; 97a, 97bl, 97bc Jason Ingram © BBC Gardeners' World Magazine; 97br Sarah Cuttle © BBC Gardeners' World Magazine; 98l Jason Ingram © BBC Gardeners' World Magazine; 98c Sarah Cuttle © BBC Gardeners' World Magazine; 98r, 99–103, 104bl, 104bc, 104br, 105a Jason Ingram © BBC Gardeners' World Magazine, 104a Matthew Taylor/Alamy Stock Photo; 105bc Andrea Koch-Link; 105br Sarah Cuttle © BBC Gardeners' World Magazine; 106l, 106c, 106r Jason Ingram © BBC Gardeners' World Magazine; 107 Tim Sandall © BBC Gardeners' World Magazine; 108a, 108bl, 108bc Jason Ingram © BBC Gardeners' World Magazine; 108br Torie Chugg © BBC Gardeners' World Magazine; 109a, 109bl Jason Ingram © BBC Gardeners' World Magazine; 109bc Tim Sandall © BBC Gardeners' World Magazine; 109br Jason Ingram © BBC Gardeners' World Magazine; 110l Torie Chugg © BBC Gardeners' World Magazine; 110c, 110r, 111, 116l, 116c, 116r Jason Ingram © BBC Gardeners' World Magazine; 117 Sarah Cuttle © BBC Gardeners' World Magazine; 118l Jason Ingram © BBC Gardeners' World Magazine; 118c Tim Sandall © BBC Gardeners' World Magazine; 118r Sarah Cuttle © BBC Gardeners' World Magazine; 119 Jason Ingram © BBC Gardeners' World Magazine; 120a raymond orton; 120bl, 120bc Jason Ingram © BBC Gardeners' World Magazine; 120br, 121a Sarah Cuttle © BBC Gardeners' World Magazine; 121bl, 121bc Jason Ingram © BBC Gardeners' World Magazine; 121br Torie Chugg © BBC Gardeners' World Magazine; 122l Jason Ingram © BBC Gardeners' World Magazine; 122c Torie Chugg © BBC Gardeners' World Magazine; 122r, 123 Jason Ingram © BBC Gardeners' World Magazine; 124a Peter Turner Photography; 124bl Jason Ingram © BBC Gardeners' World Magazine; 124bc Torie Chugg © BBC Gardeners' World Magazine; 124br Jason Ingram © BBC Gardeners' World Magazine; 125a Joe Kuis;125bl Jason Ingram © BBC Gardeners' World Magazine;125bc, 125br Sarah Cuttle © BBC Gardeners' World Magazine;126l Jason Ingram © BBC Gardeners' World Magazine; 126c, 126r Sarah Cuttle © BBC Gardeners' World Magazine; 127, 128l Jason Ingram © BBC Gardeners' World Magazine; 128c Yui Yuize; 128r, 129, 130l, 130c Jason Ingram © BBC Gardeners' World Magazine; 130r Sarah Cuttle © BBC Gardeners' World Magazine; 131 Neil Hepworth © BBC Gardeners' World Magazine; 132l, 132c Jason Ingram © BBC Gardeners' World Magazine; 132r Sarah Cuttle © BBC Gardeners' World Magazine; 133 Jason Ingram © BBC Gardeners' World Magazine; 134a Wiert nieuman; 134bl Sarah Cuttle © BBC Gardeners' World Magazine; 134bc, 134br, 135a, 135bl Jason Ingram © BBC Gardeners' World Magazine; 135bc Sarah Cuttle © BBC Gardeners' World Magazine; 135br Jason Ingram © BBC Gardeners' World Magazine; 136l Lois GoBe; 136c Sarah Cuttle © BBC Gardeners' World Magazine; 136r Jason Ingram © BBC Gardeners' World Magazine; 137 Torie Chugg © BBC Gardeners' World Magazine; 138a, 138bl, 138bc Jason Ingram © BBC Gardeners' World Magazine; 138br Edita Medeina; 139a, 139bl, 139bc, 139br, 140l Jason Ingram © BBC Gardeners' World Magazine; 140c Tim Sandall © BBC Gardeners' World Magazine; 140r Flower_Garden; 141 Jason Ingram © BBC Gardeners' World Magazine; 142l Sarah Cuttle © BBC Gardeners' World Magazine; 142c Torie Chugg © BBC Gardeners' World Magazine; 142r Sarah Cuttle © BBC Gardeners' World Magazine; 143 Jason Ingram © BBC Gardeners' World Magazine; 148l Torie Chugg © BBC Gardeners' World Magazine;148c Jason Ingram © BBC Gardeners' World Magazine; 148r, 149 ©

Liz Potter; 150a Adam Pasco © BBC Gardeners' World Magazine; 150bl Jason Ingram © BBC Gardeners' World Magazine; 150bc Joe Kuis; 150br Sarah Cuttle © BBC Gardeners' World Magazine; 151a, 151bl, 151bc, 151br, 152l, 152c Jason Ingram © BBC Gardeners' World Magazine; 152r Adam Pasco © BBC Gardeners' World Magazine; 153, 154l Jason Ingram © BBC Gardeners' World Magazine; 154c Torie Chugg © BBC Gardeners' World Magazine; 154r, 155, 156l, 156c, 156r Jason Ingram © BBC Gardeners' World Magazine; 157 © Liz Potter; 158a, 158bl, 158bc, 158br, 159a, 159bl Jason Ingram © BBC Gardeners' World Magazine; 159bc Sarah Cuttle © BBC Gardeners' World Magazine; 159br Torie Chugg © BBC Gardeners' World Magazine; 160l Jason Ingram © BBC Gardeners' World Magazine; 160c Adam Pasco © BBC Gardeners' World Magazine; 160r, 161 Jason Ingram © BBC Gardeners' World Magazine; 162l Paul Debois © BBC Gardeners' World Magazine; 162c, 162r, 163 Jason Ingram © BBC Gardeners' World Magazine; 164l Paul Debois © BBC Gardeners' World Magazine; 164c, 164r Sarah Cuttle © BBC Gardeners' World Magazine; 165 Paul Debois © BBC Gardeners' World Magazine; 166l Jason Ingram © BBC Gardeners' World Magazine; 166c, 166r Jason Ingram © BBC Gardeners' World Magazine; 167, 168l, 168c Paul Debois © BBC Gardeners' World Magazine; 168r Sarah Cuttle © BBC Gardeners' World Magazine; 169 Paul Debois © BBC Gardeners' World Magazine; 170a N.Stertz; 170bl Paul Debois © BBC Gardeners' World Magazine; 170bc Jason Ingram © BBC Gardeners' World Magazine; 170br Torie Chugg © BBC Gardeners' World Magazine; 171a, 171bl Jason Ingram © BBC Gardeners' World Magazine; 171bc Tim Sandall © BBC Gardeners' World Magazine; 171br Sarah Cuttle © BBC Gardeners' World Magazine; 172l Jason Ingram © BBC Gardeners' World Magazine; 172c, 172r Sarah Cuttle © BBC Gardeners' World Magazine; 173 Paul Debois © BBC Gardeners' World Magazine; 174a, 174bl Sarah Cuttle © BBC Gardeners' World Magazine; 174bc, 174br Jason Ingram © BBC Gardeners' World Magazine; 175a © Liz Potter; 175bl Sarah Cuttle © BBC Gardeners' World Magazine; 175bc Jason Ingram © BBC Gardeners' World Magazine; 175br © Liz Potter; 180l Jason Ingram © BBC Gardeners' World Magazine; 180c Tim Sandall © BBC Gardeners' World Magazine; 180r, 181 Jason Ingram © BBC Gardeners' World Magazine; 182l Sarah Cuttle © BBC Gardeners' World Magazine; 182c Jason Ingram © BBC Gardeners' World Magazine; 182r, 183 Paul Debois © BBC Gardeners' World Magazine; 184l, 184c Jason Ingram © BBC Gardeners' World Magazine; 184r Sarah Cuttle © BBC Gardeners' World Magazine; 185, 186l Jason Ingram © BBC Gardeners' World Magazine; 186c, 186r, 187 Sarah Cuttle © BBC Gardeners' World Magazine; 188a Joe Kuis; 188bl, 188bc Sarah Cuttle © BBC Gardeners' World Magazine; 188br Jason Ingram © BBC Gardeners' World Magazine; 189a Paul Debois © BBC Gardeners' World Magazine; 189bl, 189bc, 189br Jason Ingram © BBC Gardeners' World Magazine; 190l Sarah Cuttle © BBC Gardeners' World Magazine; 190c, 190r Torie Chugg © BBC Gardeners' World Magazine; 191, 192l Sarah Cuttle © BBC Gardeners' World Magazine; 192c Jason Ingram © BBC Gardeners' World Magazine; 192r Paul Debois © BBC Gardeners' World Magazine; 193 Sarah Cuttle © BBC Gardeners' World Magazine; 194a Gina Kelly/Alamy Stock Photo; 194bl, 194bc Jason Ingram © BBC Gardeners' World Magazine; 194br © Liz Potter; 195a Jason Ingram © BBC Gardeners' World Magazine; 195bl Paul Debois © BBC Gardeners' World Magazine; 195bc Jason Ingram © BBC Gardeners' World Magazine; 195br Sarah Cuttle © BBC Gardeners' World Magazine; 196l, 196c Jason Ingram © BBC Gardeners' World Magazine;196r Walter Erhardt; 197, 198l Jason Ingram © BBC Gardeners' World Magazine; 198c Paul Debois © BBC Gardeners' World Magazine; 198r Torie Chugg © BBC Gardeners' World Magazine; 199 Sarah Cuttle © BBC Gardeners' World Magazine; 200l Paul Debois © BBC Gardeners' World Magazine; 200c Jason Ingram © BBC Gardeners' World Magazine; 200r Torie Chugg © BBC Gardeners' World Magazine; 201 Jason Ingram © BBC Gardeners' World Magazine; 202l Paul Debois © BBC Gardeners' World Magazine; 202c Nancy J. Ondra; 202r Sarah Cuttle © BBC Gardeners' World Magazine; 203 Paul Debois © BBC Gardeners' World Magazine; 204l, 204c, 204r, 205, 206l Jason Ingram © BBC Gardeners' World Magazine; 206c Paul Debois © BBC Gardeners' World Magazine; 206r Jason Ingram © BBC Gardeners' World Magazine; 207 Paul Debois © BBC Gardeners' World Magazine; 208l, 208c, 208r, 209 Sarah Cuttle © BBC Gardeners' World Magazine; 210a Tim Sandall © BBC Gardeners' World Magazine; 210bl Jason Ingram © BBC Gardeners' World Magazine; 210bc Sarah Cuttle © BBC Gardeners' World Magazine; 210br Jason Ingram © BBC Gardeners' World Magazine; 211a freya-photographer; 211bl, 211bc Sarah Cuttle © BBC Gardeners' World Magazine; 211br Jason Ingram © BBC Gardeners' World Magazine; 216l Sarah Cuttle © BBC Gardeners' World Magazine; 216c Jason Ingram © BBC Gardeners' World Magazine; 216r Torie Chugg © BBC Gardeners' World Magazine; 217 aniana; 218l Paul Debois © BBC Gardeners' World Magazine; 218c, 218r Sarah Cuttle © BBC Gardeners' World Magazine; 219 Jason Ingram © BBC Gardeners' World Magazine; 220l Sarah Cuttle © BBC Gardeners' World Magazine; 220c Jason Ingram © BBC Gardeners' World Magazine; 220r © Liz Potter; 221, 222–223 Jason Ingram © BBC Gardeners' World Magazine; 224l, 224c Paul Debois © BBC Gardeners' World Magazine; 224r, 225 Jason Ingram © BBC Gardeners' World Magazine; 226l Sarah Cuttle © BBC Gardeners' World Magazine; 226c Orest lyzhechka; 226r © Liz Potter; 227 Sarah Cuttle © BBC Gardeners' World Magazine; 228l Paul Debois © BBC Gardeners' World Magazine; 228c Jason Ingram © BBC Gardeners' World Magazine; 228r Paul Debois © BBC Gardeners' World Magazine; 229 Jason Ingram © BBC Gardeners' World Magazine; 230l Torie Chugg © BBC Gardeners' World Magazine; 230c Jason Ingram © BBC Gardeners' World Magazine; 230r Paul Debois © BBC Gardeners' World Magazine; 231 Sarah Cuttle © BBC Gardeners' World Magazine; 232a, 232bl, 232bc Jason Ingram © BBC Gardeners' World Magazine; 232br Paul Debois © BBC Gardeners' World Magazine; 233a Susan & Allan Parker/Alamy Stock Photo; 233bl Jason Ingram © BBC Gardeners' World Magazine; 233bc Sarah Cuttle © BBC Gardeners' World Magazine; 233br Torie Chugg © BBC Gardeners' World Magazine; 234l Avalon.red/Alamy Stock Photo; 234c, 234r, 235, 236l, 236c Jason Ingram © BBC Gardeners' World Magazine; 236r Sarah Cuttle © BBC Gardeners' World Magazine; 237 Jason Ingram © BBC Gardeners' World Magazine; 238l Torie Chugg © BBC Gardeners' World Magazine; 238c Paul Debois © BBC Gardeners' World Magazine; 238r, 239, 240l, 240c Jason Ingram © BBC Gardeners' World Magazine; 240r JohnatAPW; 241, 242a, 242bl Jason Ingram © BBC Gardeners' World Magazine; 242bc Tim Sandall © BBC Gardeners' World Magazine; 242br, 243a Jason Ingram © BBC Gardeners' World Magazine; 243bl Tim Sandall © BBC Gardeners' World Magazine; 243bc Paul Debois © BBC Gardeners' World Magazine; 243br, 248l, 248c, 248r, 249, 250l Jason Ingram © BBC Gardeners' World Magazine; 250c, 250r, 251 Sarah Cuttle © BBC Gardeners' World Magazine; 252a Paul Debois © BBC Gardeners' World Magazine; 252bl Sarah Cuttle © BBC Gardeners' World Magazine; 252bc, 252br, 253a, 253bl Jason Ingram © BBC Gardeners' World Magazine; 253bc Sarah Cuttle © BBC Gardeners' World Magazine; 253br, 254l Jason Ingram © BBC Gardeners' World Magazine; 254c Torie Chugg © BBC Gardeners' World Magazine; 254r Jason Ingram © BBC Gardeners' World Magazine; 255, 256l Sarah Cuttle © BBC Gardeners' World Magazine; 256c Torie Chugg © BBC Gardeners' World Magazine; 256r, 257, 258l Jason Ingram © BBC Gardeners' World Magazine; 258c Tom Meaker; 258r Sarah Cuttle © BBC Gardeners' World Magazine; 259, 260l Jason Grant; 260c Paul Debois © BBC Gardeners' World Magazine; 260r Sarah Cuttle © BBC Gardeners' World Magazine; 261, 262a Jason Ingram © BBC Gardeners' World Magazine; 262bl, 262bc Paul Debois © BBC Gardeners' World Magazine; 262br © Liz Potter; 263a, 263bl, 263bc Jason Ingram © BBC Gardeners' World Magazine; 263br Mei Yi; 264l, 264c Jason Ingram © BBC Gardeners' World Magazine; 264r Paul Debois © BBC Gardeners' World Magazine; 265 Sarah Cuttle © BBC Gardeners' World Magazine; 266l Jason Ingram © BBC Gardeners' World Magazine; 266c, 266r, 267 Sarah Cuttle © BBC Gardeners' World Magazine; 268a, 268bl Paul Debois © BBC Gardeners' World Magazine; 268bc, 268br, 269–271 Jason Ingram © BBC Gardeners' World Magazine; 272l Torie Chugg © BBC Gardeners' World Magazine; 272c Neil Hepworth © BBC Gardeners' World Magazine; 272r Jason Ingram © BBC Gardeners' World Magazine; 273 Torie Chugg © BBC Gardeners' World Magazine; 274a, 274bl, 274bc Sarah Cuttle © BBC Gardeners' World Magazine; 274br Jason Ingram © BBC Gardeners' World Magazine; 275a Bayhu19; 275bl, 275bc Sarah Cuttle © BBC Gardeners' World Magazine; 275br, 276l, 276c Jason Ingram © BBC Gardeners' World Magazine; 276r Nancy J. Ondra; 277, 278l Jason Ingram © BBC Gardeners' World Magazine; 278c Sarah Cuttle © BBC Gardeners' World Magazine; 278r, 279 Jason Ingram © BBC Gardeners' World Magazine;

Index

Note: page numbers in **bold** refer to illustrations.

D

E

F

G

H

I

L

M